British Independent Airlines since 1946

by

A. C. MERTON JONES

•

A comprehensive history of the operations of all independent airlines and air taxi operators who have played some part, however small, in the development of the post-war private airline industry of the United Kingdom and the Republic of Ireland.

"This book is dedicated to Marigold, and to the many people who through their hard work, enterprise and enthusiasm over the last thirty years, have provided me with such a fascinating subject to write about".

•

PUBLISHED JOINTLY BY LAAS INTERNATIONAL, 21 BARCHESTER CLOSE, UXBRIDGE UB8 2JY

and

MERSEYSIDE AVIATION SOCIETY LTD., LIVERPOOL AIRPORT, LIVERPOOL L24 8QE

Volume 1 of four
ISBN 0 902420 07·0

PREFACE

This book is the result of many years of research into the ever fascinating world of British air transport since the lifting of the ban on private flying on January 1 1946. Some time ago Mr.D.K.Fox wrote a series of interesting histories of British independent airlines in the fortnightly Air Britain publication known as British Civil Aviation News. In order to help illustrate the complexity of compiling a work of this size I would like to quote from Mr.Fox's introductory passage from the British Civil Aviation News for October 25 1958, in which he says - 'If the whole story is ever written, it will be found to be a sizeable volume, and one of quite extraordinary interest. It is unlikely to be written in full, for that would involve exposing the conduct not only of some operators, but of the lunatic fringe of politicians (and their sponsors) who have endeavoured to disrupt the industry'. While not claiming to have written a complete history of our postwar private airline industry, I do hope that this book will go some way to filling a gap in aviation history, which has so far been left unrecorded.

The information contained in the histories and fleet lists has come from a very large number of varying sources. Many details in this book, particularly those contained in the fleet lists, will contradict earlier information published in other books and magazines. To help to illustrate this point, I would like to take as an example the acquisition of the Trans Canada Airlines' North Star fleet by Overseas Aviation in 1961. Delivery dates of these aircraft to Gatwick in this book will contradict dates published some years ago by an aviation group. The dates contained in this book, for these aircraft, have been obtained from official sources and having examined these documents at first hand, I have no doubt as to their accuracy.

When starting to type the fleet lists, it became clear that any excessively long constructor's number would use up too much valuable space. As these constructor's numbers are almost always those of Piper light aircraft, it was decided to shorten these numbers by the removal of the date of construction, which is placed after the aircraft's type number. For example a Piper Aztec G-BBCU c/n 27-7305112, becomes c/n 27-5112 - thus losing the fact that the aircraft was a 1973 model. Piper aircraft built before 1972 are unaffected by this, as the year of the model was not included prior to this date. Another problem has been the variation in spelling of several Continental cities, and any unusual spelling of a foreign city is generally due to the fact that that city has had two or more variations in spellings since the last war.

The fleet lists of each company follow a brief history of each airline, and in these fleet lists I have tried to include as much relevant information as possible. The aircraft are listed by aircraft type, followed by its registration, constructor's number, delivery, registration or purchase date, sale or disposal date, and then any remarks about the aircraft which are necessary and relevant to the airline. Where possible, delivery dates are included for the aircraft, but where no delivery date is obtainable, I have substituted a sale or registration date to that airline. Unfortunately, because of the size of this work, some errors, particularly in the fleet lists, must have crept in, and although I have examined all dates as thoroughly as possible, mistakes will inevitably be present in the fleet lists. One of the major problems facing me was the incredible variation in dates quoted by different publications for the same aircraft - the worst nightmare in this respect has been the construction of an accurate fleet list for Eagle Aviation's Vikings, and I really cannot guarantee that list's complete accuracy. However, with the mass of information available to me, I hope that the end product is presentable and will bring back memories, happy I hope, of those golden days now sadly gone by, to the people who take the trouble to read this work.

Obviously, it would have been impossible to have put all this information together alone, and I would like to thank the very many people who have given so freely of their time in helping me to complete this work. Ian MacFarlane has for many years played a very important part as a correspondent and photographer for the Newcastle area, recording in detail all the events that have taken place at Woolsington. His never ceasing enthusiasm and help which he has given me has, on many occasions, shamed me into spending more and more time on the book. Ian has supplied a vast selection of top quality prints of aircraft for use in this book, and his pestering of pilots and aircrew at Newcastle has led to a fund of interesting information coming to light. I hope that Ian's natural modesty will allow him to accept my very greatest thanks for all his wonderful assistance. Another corner of the country where aviation history has been well recorded over the last thirty years is Southend, and from here I was lucky to be able to use prints from the collections of Mr.A.J.Jackson and Mr.John Goring. Mr.A.J.Jackson needs no introduction as one of Britain's foremost aviation historians, and his time and patience in searching through his vast negative collection and then producing a most beautiful selection of prints, has meant that many more of the early post-war airlines have been illustrated than would have otherwise been. John Goring's interesting collection of photographs contains some of the more unusual aircraft of the sixties, and many of the more exciting photographs in this book are also from his collection. From the south west of England I have been fortunate enough to have been able to obtain a varied selection of high quality photographs from Austin J.Brown. Austin's superbly clear and sharp photographs have been a delight to include in this book, and it is perhaps rather unfortunate that most are only of postcard size. Mr.Brian Stainer of Aviation Photo News has also supplied me with a selection of his high quality photographs for use in this book, and many of his prints have filled in gaps in the photographic coverage.

I would like to have said something about all the contributors for this book but, sadly, space limits me to just mentioning their names and saying one big 'thank you'. Other photographs have come from the collections of Mr.W.Stevenson, Mr.Stan Davidson, Mr.Arthur Pearcy Jnr., Mr.Peter J.Bish, Tony Large, Gordon Bain, Phil Butler and Michael Stroud. In addition, Stephen Piercey has spent many hours delving through the negative files of 'Flight International', and has produced many fascinating photographs. Dave Dorman has obtained many photographs of BAC 1-11s, VC10s, Viscounts and Vikings from the British Aircraft Corporation, while the colour covers have been produced from transparancies kindly supplied by Mr.Philip Birtles of Hawker Siddeley Aviation, the collection of Gordon Bain and the British Aircraft Corporation.

Many of the people already mentioned have also supplied me with details for the book in the form of their notes, newspaper cuttings, books, magazines and official documents. Phil Butler and Keith Crowden of the Merseyside Aviation Society have been very helpful in this respect, and Derek King, Tony Eastwood, Stan Wilkinson, Dave Dorman, Dave Skinner and Raymond Hoddinott of LAAS International have also been particularly helpful. Many airlines have sent me details and photographs for inclusion, and I would like to say a special word of thanks to Mr.S.K.Palmer of Cabair, Mr.Peter Moxham of Thurston Aviation, Mr.B.L.Williams of Air London, Mrs.L.M.Wilcox of British Air Ferries/Transmeridian Air Cargo, Mr.Trevor Kavaney of Britannia Airways, Mr.Julian Rogers of Intra Airways and Captain B.G.Cramp of British Midland Airways. Other essential details have flowed from the pens of Dave Bourgourd in Guernsey, and the many people who wrote to me after the publication of an earlier edition of this book.

People who do not very often receive thanks for their help are the staff at the many airports around the country, who allow people like myself access to the manoeuvring area of an airfield for the purpose of taking photographs. While some were not especially helpful, I would like to say thank you to the air traffic control, customs and immigration, and security staffs at Aberdeen, Blackbushe, Coventry, Jersey, Lydd, Southend and Staverton Airports for their assistance.

Finally, to Tony Eastwood who has typed this book, I would like to express my gratitude for his patience and devotion to this marathon task. With all this excellent typing practice, coupled with the fact that men now have equal rights with women, I am sure that someone somewhere must require the services of a typist of his qualities! Perhaps I should also take this opportunity to thank his wife for her patience in living with a person glued to a typewriter day and night.

By way of an introduction to this book, and to try to get people into the right mood for reading this book, I have written a brief resume of the first ten years of British civil aviation since the last war. I hope that if readers start with this introduction, then it may help them to visualise the operations of airlines described later on in the book.

January 1 1976 — Tony Merton Jones, Maidenhead.

BIBLIOGRAPHY

Air Britain	The Douglas Commercial Story; the Douglas DC4; the Douglas DC6/7; the Lockheed Constellation; Fleet Operators; the United Kingdom and Eire Civil Registers; Business Jets International; the DH Dove and Heron; the Viking, Valetta and Varsity.
Airline Publications & Sales	The Handley Page Herald, the Short Skyvan.
Allmark Publications	The Avro Anson.
Blandford	Civil Airliners since 1946.
Cassell	Berlin Airlift by Robert Rodrigo (1960).
Hodder & Stoughton	Airway to the Isles by Philip Cliefe (1966).
Ian Allan	British Civil Aircraft Register; Civil Aircraft Markings; Worlds Air Fleets; the Handbook of the Vickers Viscount.
LAAS International	The Douglas DC3; Vickers Turbine Transports; the BAC 1-11.
Merseyside Aviation Society	Irish Aircraft; the Short Skyvan.
Midland Counties Aviation Society	Beagle Aircraft.
Profile Publications	The Avro York; the Dragon Rapide; the DH Comet; the Vickers Viscount 700.
Putnams	British Civil Aircraft since 1919 by A.J.Jackson, volumes 1 - 3; Bristol Aircraft since 1910; Avro Aircraft since 1908; Airspeed Aircraft since 1931; Miles Aircraft since 1925; European Transport Aircraft since 1910.
Souvenir Press	The Sky Tramps by Peter Jackson (1965).
Temple Press	The Aeroplane Directory of British Aviation.

The following periodicals:-

Flight International; the Aeroplane & Commercial Aviation News; Air Pictorial; Pilot & Light Aeroplane; Air Enthusiast; Air International; Flying Review International; Aircraft Illustrated; Aircraft Illustrated Extra; the Aeroplane Spotter; Aviation News; Popular Flying; Interavia; Aeroplane Monthly; Aeronautics; International Freighting Weekly; and the Travel Trade Gazette.

The following enthusiast's publications:-

Air Britain 'Digest', 'British Civil Aviation News' and 'British Civil Register News'; Anglian Aviation Society 'Anglia Aeronews'; Air Circle International 'Circuit 74'; British Aircraft Preservation Council 'Control Column'; Central Scotland Aviation Group 'Scottish Air News'; Air North 'Woolsington Movements'; Gatwick Aviation Society 'Hawkeye'; LAAS International 'Aviation News & Review'; Midland Counties Aviation Society 'Airstrip'; Merseyside Aviation Society 'Flypast'; South West Aviation Society 'South West Aviation News'; West London Aviation Group 'South East Air Review'; Prestwick Air Letter'; 'Home Counties Aeronews'; 'Scotland Scanned'; 'Airfleet'; 'Aero Review Holland'; 'Aviation Letter'; 'Yorkshire Air News'; 'Southend Aeronews'; 'Irish Air Letter'; and various newsletters including those of the Gwent Branch of Air Britain, and the Kent Branch of Air Britain.

ABBREVIATIONS

Explanation of abbreviations used in text:-

AOC	Air Operator's Certificate	bt.	Date of purchase by operator
ATAC	Air Transport Advisory Council	ca.	Initial Certificate of Airworthiness date
ATLB	Air Transport Licensing Board	c/n	Constructor's number
BoT	Board of Trade	dd.	delivery date to operator
CAA	Civil Aviation Authority	lsd.	delivery date to operator on lease
ISOC	In service on closure of airline	rd.	registration date to operator
ISOF	In service on airline's formation	ret.	date of return of aircraft off lease to owner
ISOI	In service on airline's integration with parent airline	wfu.	withdrawn from use date
ISOM	In service on airline's merger	C of A	Certificate of Airworthiness
ISOT	In service on airline's take-over		

PHOTO CAPTIONS

FRONT COVER	British Eagle BAC 1-11	(B.A.C.)
BACK COVER	British Island Airways Herald	(Gordon Bain)
	Aer Lingus Boeing 737	(Gordon Bain)

1945 - 1955 THE FIRST TEN YEARS

When the War with Germany ended in the summer of 1945, civil aviation in the British Isles consisted of a few wartime or pre-1939 aircraft flying in the colours of a handful of British airlines including British Overseas Airways Corporation (solely operating international services), Railway Air Services, Jersey Airways, Scottish Airways, Air Commerce, Allied Airways, Isle of Man Air Services, Olley Air Service, Great Western and Southern Airlines and West Coast Air Services. In addition, Aer Lingus was flying scheduled services in conjunction with West Coast Air Services between Dublin and Liverpool. All these airlines were flying vintage de Havilland biplanes of varying marks on scheduled internal air services. During the war years, these airlines flew these services with amazing efficiency taking into account the shortage of both aircraft and spares. Until the opening of the Liverpool to London service in November 1944, all these schedules had required a sea crossing and by August 1945 the following scheduled services were in operation:-

Air Commerce, Speke, flying Rapides (G-AERZ, G-AFEP) Liverpool to Belfast, London and the Isle of Man.

Allied Airways, Dyce, flying Dragon (G-ACNJ) and Rapides (G-ACZE, G-ACZF, G-ADAH, G-AGDM, G-AGHI) Aberdeen - Wick - Kirkwall - Shetland.

Great Western & Southern Airlines, Land's End and Speke, flying Rapides (G-ACPP, G-AGEE) Liverpool to London and Belfast; Land's End to the Scilly Isles.

Isle of Man Air Services, Speke, flying Rapides (G-AFEZ, G-AGSJ) Liverpool to the Isle of Man.

Jersey Airways, Jersey, flying Rapides (G-AGPH, G-AGPI, G-AGSH, G-AGSK) Jersey to Guernsey and London; Guernsey to London.

Olley Air Service, Speke, flying Rapide (G-AGSI) on charter flights only.

Railway Air Services, Croydon and Renfrew, flying Rapides (G-AFFF, G-AGLE, G-AGLP, G-AGLR) and Dragon Expresses (G-ACVY, G-ACZP, G-AENR) London - Liverpool - Belfast; Liverpool - Dublin (DH86s); Glasgow - Belfast; Glasgow - Liverpool - London (DH89s).

Scottish Airways, Inverness and Renfrew, flying Dragon (G-ACIT) and Rapides (G-ADAJ, G-AEWL, G-AFOI, G-AFRK, G-AGDG, G-AGIC, G-AGIF, G-AGJF, G-AGJG) Inverness - Kirkwall - Shetland; Inverness - Stornoway; Glasgow - Campbeltown; Glasgow - Islay; Glasgow - Tiree - Benbecula - North Uist - Stornoway; Glasgow - Stornoway; Glasgow - Belfast.

West Coast Air Services, Speke, flying Dragon Express (G-ADYH) and Rapide (G-AERN) Liverpool-Dublin (DH86); Liverpool - Belfast; Liverpool - London (DH89).

All these aircraft, with the exception of those operated by Allied Airways, came under the authority of the Associated Airways Joint Committee Group which had been set up to co-ordinate and organise a network of domestic scheduled services by these airlines during the War.

Before the end of 1945, the first Avro XIXs were appearing on some of these routes. Railway Air Services had ordered a fleet of this new civil derivative of the Anson navigation trainer and communications hack, and one of these aircraft opened a new scheduled service between London and Belfast on December 3 1945. During December it was announced by the new Government that the private flying ban would be lifted on December 31 1945, allowing private operators to start charter flying once again.

The new Government also put a ban on any new scheduled services by these new carriers, as it intended that the newly formed British European Airways Corporation (B.E.A.C) should operate all internal scheduled services as soon as that airline could obtain enough suitable crews and aircraft for these services. However on January 1 1946, civil aviation got off to a brisk start - an Avro Lancastrian of British South American Airways (G-AGWG) was the first aircraft to take off from Heathrow Airport on January 1. Named 'Star Light', this aircraft left on a proving flight to Buenos Aires via Lisbon, Bathurst, Natal, Rio de Janeiro and Montevideo. The first civil flying training school to reopen on January 1 was Marshall's Flying School at Cambridge operating a fleet of Tiger Moths. Also on New Year's Day, Hunting Air Travel flew the first post war air taxi flight in Britain with a Percival Proctor (G-AGSX), and the first private charter flight was made by a Taylorcraft Auster J/1 of Cambrian Air Services. This aircraft, piloted by Mr. Eric Symmons, flew a load of aircraft parts from Cardiff to Filton on January 1.

During the next few months, numerous air charter firms sprang up including Morton Air Services who initially operated two Rapides on charter work from Croydon. The company's first charter took place on January 21 1946 - a flight to Zurich using a Rapide specially equipped with removable long range tanks increasing the range to a maximum of 650 miles. Amongst other companies to start at this time was Silver City Airways who operated a fleet of Avro Lancastrians on long range charters from Dunsfold. One of these aircraft was the first aircraft to make an overseas flight from New Zealand's South Island, when G-AHBW (City of London) left Harewood Aerodrome, Christchurch for Melbourne Airport, Australia. British American Air Services took delivery on August 25 1946 of a Bristol Freighter for a passenger /cargo air charter service between London and Nice. There were, of course, many other new airlines too, and with an abundance of surplus Royal Air Force aircraft flooding onto the second-hand aircraft market, airlines were being formed in every part of the country.

These aircraft included Percival Proctors, Q.6s, Dakotas, Ansons, Oxfords, Dominies, Halifaxes, Lancasters, numerous other light de Havilland types and many American aircraft. Some aircraft manufacturing companies bought these aircraft, converted them, and then resold them to civil operators. The Airspeed Company of Christchurch bought large numbers of their Oxford trainer from the Royal Air Force, and converted them into good air taxi and charter aircraft known as the Consul. Many were exported, but a large number flew with British airlines on charter and tourist services carrying up to eight passengers in relative comfort for that time. Short Brothers converted a number of Halifax bombers into freighters, and these served for a short time with some of the new post war airlines, but most had been withdrawn by the end of the Berlin Airlift. Airlines themselves converted aircraft - Scottish Aviation became the Douglas appointed United Kingdom agent for the conversion of Dakotas to civilian DC 3 standard. In addition, Field Aircraft Services at Tollerton, and Skyways at Dunsfold, converted many Dakotas, and Western Airways at Locking and Whitchurch Airfields converted a variety of wartime aircraft including Ansons, Arguses, Percival Proctors, Q.6s and Piper Cubs.

Many companies were formed by one man spending his entire savings on the purchase of the aircraft - or by borrowing the money from numerous sources. East Anglian Flying Services was formed by Squadron Leader R.J. (Jack) Jones when he spent his retirement gratuity from the Royal Air Force on an Airspeed Courier and a Puss Moth - both of which he bought at one of the many post war surplus Royal Air Force aircraft sales at Kemble. Another airline - Skytravel - was formed after the owner had leased his own private aircraft out, and finding this to be remarkably profitable he formed his own airline. With the end of the War still so near, and the spirit of aviation still very much alive in people, many operators made money out of pleasure flying. With public enclosures set up at many aerodromes all over the country including Heathrow, Northolt, Whitchurch, Cardiff and Weston, many companies would start flying in the morning and continue flying passengers until dusk.

The first post-war British air race was held at Lympne Airfield on Saturday August 31 and Sunday September 1 1946. The Folkestone Trophy Race was flown in three heats over three laps of the old pre-war air racing course. The turning points were the Capel Airship Shed, the Folkestone Harbour light, the Hythe gasometer and finally the centre of Lympne Airfield. The first heat was won by John Grierson flying a Supermarine Walrus (G-AHFN); the second heat by Mr.S.T.Carruthers in a Percival Gull Six (G-ADPR); and the third heat by Mr D.M.Bay in a Percival Proctor 1 (G-AHMV). In the final on the Sunday, the Walrus led from the end of the first lap until the finish, to win at an average speed of 121 mph.

On Monday September 2 1946, the first regular internal air service started from Northolt, when Railway Air

Services opened a direct Northolt to Renfrew service, replacing the Croydon - Prestwick - Renfrew run. On the same day, Scottish Airways (under charter to BEAC), commenced a direct Northolt to Prestwick scheduled service. On October 7 1946, B.E.A.C commenced its winter schedules, but pending the formation of the English and Scottish Divisions of the Corporation, these services were operated on a three monthly agreement by their former owners. The winter schedules were as follows:-

ENGLISH DIVISION

Railway Air Services

1. London - Liverpool - Belfast, daily except Sundays by DH89A.
2. Liverpool - Belfast, twice daily except Sundays by DH89A.
3. London - Belfast, twice daily except Sundays by Avro X1X.
4. London - Manchester - Belfast, twice daily except Sundays by Avro X1X.

Isle of Man Air Services

5. Liverpool - Isle of Man, five times daily except Sundays by DH89A.
6. Isle of Man - Blackpool - Manchester, twice daily except Sundays by DH89A.
7. Isle of Man - Carlisle, Mondays, Wednesdays and Fridays by DH89A.

Great Western & Southern Airlines

8. Bristol - Southampton, twice daily except Sundays by DH89A.
9. Bristol - Cardiff, four times daily by DH89A.
10. Penzance - Scilly Isles, four times daily except Sundays by DH89A.
11. Cardiff - Weston, eight times daily by DH89A.

Channel Islands Airways

12. Guernsey - Alderney, twice daily on Mondays, Tuesdays, Fridays and Saturdays by DH89A.
13. Jersey - Guernsey, four times daily by DH89A.
14. Southampton - Guernsey, three times daily by DH89A.
15. Jersey - Southampton, three times daily by DH89A.
16. London - Guernsey, daily by DH89A.
17. London - Jersey, twice daily by B.170, once on Wednesday by DH89A.

SCOTTISH DIVISION

Railway Air Services / Scottish Airways

1. Glasgow - London, daily except Sundays by DC 3.

Scottish Airways

2. Glasgow - Stornoway, daily except Sundays by DH89A.
3. Glasgow - Hebrides, daily except Sundays by DH89A.
4. Glasgow - Campbeltown, daily except Sundays by DH89A.
5. Glasgow - Islay, daily except Sundays by DH89A.
6. Glasgow - Belfast, daily except Sundays by DC 3.
7. Glasgow - Orkney - Shetland, daily except Sundays by DH89A.
8. Aberdeen - Inverness - Stornoway, first stage daily except Sundays, second stage on Monday, Wednesday and Friday by DH89A.
9. Inverness - Wick - Orkney, daily except Sundays by DH89A.
10. Kirkwall - Trans Pentland, no set schedule by DH89A.
11. Prestwick - Belfast, daily except Sundays by DC 3.

Allied Airways

12. Aberdeen - Shetland, daily except Sundays by DH89A.
13. Aberdeen - Wick - Orkney - Shetland, daily except Sundays by DH89A.
14. Aberdeen - Orkney, Monday, Tuesday, Friday and Saturday by DH89A.

And so 1946 was drawing to a close with the darkening shadow of B.E.A.C looming over the few British independent scheduled airline companies. The charter companies, though, were flourishing and some of the larger companies were able to order new equipment. Hunting Air Travel ordered a fleet of new twenty one seat Vickers Vikings for use on its charter passenger services to Europe, and many operators were flying Bristol Freighters, including Airwork and British American Air Services. The majority of the surplus Royal Air Force aircraft suffered from poor performances when fully loaded, or when engine failure occurred, and the Bristol Freighter and Viking provided the airlines with a better all round aircraft both in load carrying capabilities and in performance in the event of power failure. By the end of 1946, new aircraft were either already in service or in the design stage and these included the Ambassador from Airspeeds and the Apollo from Armstrong Whitworth Company, The Lancastrian and Tudor from Avros, the Freighter/Wayfarer and Brabazon from Bristols, the Concordia from Cunliffe Owen, the Dove and Comet from de Havillands, the Hermes from Handley Page, The Aerovan and Marathon from Miles, and the Viking and Viceroy (later known as Viscount) from Vickers.

On February 1 1947, B.E.A.C was at last ready to take over most of the private airlines' scheduled services and their aircraft, and so on this date all those companies controlled by the Associated Airways Joint Committee (AAJC) with two exceptions, passed into the hands of B.E.A.C , Railway Air Services (RAS) handed over a fleet of two DC 3s, and several Dragon Rapides and Avro X1Xs; in addition a number of Junkers Ju52/3ms were also handed over, but these aircraft had not been used by R.A.S. for passenger carrying, and only one, (G-AHBP), had been flown on cargo services under the class name of 'Jupiter'. Scottish Airways' fleet of Dragon Rapides and one Dragon had been supplemented by two Junkers Ju52s (G-AHOD and G-AHOK), but although these aircraft arrived at Renfrew before the takeover, they never entered service with Scottish. By the time of the handover, Isle of Man Air Services had opened a thrice weekly service from the Isle of Man to Glasgow, and for many of its services had wet leased Railway Air Services' Avro X1Xs when loads were heavy, but on February 1, a fleet of seven Dragon Rapides passed to B.E.A.C . Great Western & Southern Airlines handed over a fleet of six Dragon Rapides to B.E.A.C ., and so including a number of aircraft held by the AAJC in a pool for operation by these companies when needed, B.E.A.C. inherited a total of eight Ju52s, two DC 3s, one de Havilland Dragon, thirty nine Rapides and thirteen Avro X1Xs.

Two airlines - Channel Islands Airways and Allied Airways - continued to operate independently for a short while, but the former was taken over on April 1, and the latter on April 12 1947. B.E.A.C. divided its internal schedules into two divisions - the English Division based at Speke Airport, Liverpool, and the Scottish Division based at Renfrew Airport, Glasgow. For the future, B.E.A.C. planned to introduce a fleet of Miles Marathons and Vickers Vikings onto these services, but until these aircraft were available, further Dakotas were used.

April 12 1947 therefore marked the day of the last privately operated scheduled service, and for the time being, all British independent airlines wishing to continue flying were forced to turn to charter work. One such company, rapidly expanding its operations, was London Aero and Motor Services. Flying a fleet of converted Halifax bombers, this company operated the first British air tramp service. On April 23 1947, a Halifax, 'Port of Sydney', left Stansted on a flight which took it round the world via Canada, America, Australia, India and many smaller islands en route. The Captain for the trip was Captain Thiele, DSO., DFC., and the First Officer was the airline's Managing Director Dr. Graham Humby. During the trip, the company made several important contacts around the world, which resulted in the setting up of subsidiary companies in Sydney, Australia and

Johannesburg, South Africa. The airline's plan was to base six aircraft at Sydney and Johannesburg to carry out freight charters to any part of the world, with a further ten machines held in reserve in the United Kingdom for use at either of its two foreign bases should the need arise. The flight, which was carried out without a single hitch, involved the carriage of many unusual loads including the transport of seven tons of empty beer bottles from Suva, Fiji, to New Zealand, and also the carriage of seven tons of dripping sent from the people of Sydney to the people of Great Britain. The Halifax arrived back at Stansted on June 4 1947 to coincide with the introduction of customs facilities for the clearance of freight at Stansted.

On Wednesday May 28 1947, a Lancaster aircraft flown by Air Vice Marshall Donald Bennett left Heathrow Airport for Kindley Field, Bermuda. This was the first round trip of air refuelling trials by Flight Refuelling in conjunction with British South American Airways, and on this first flight was Sir Alan Cobham, Managing Director of Flight Refuelling. The Lancaster (G-AHJV) flew via the Azores where the refuelling took place from a second Lancaster tanker. The 4,000 miles to Bermuda took twenty hours, while the return journey took fifteen and a half hours. The aircraft received 1,800 gallons of fuel outward, and 1,600 gallons on the return journey. These trials continued until mid August 1947, and the service was operated by four specially equipped Lancasters and during the three months of operations twenty two separate crossings were made. Fare-paying passengers and freight were not carried. Of the twenty two flights made, twenty were completely successful and only one partly successful. The system of interception was in five phases, and the actual refuelling was carried out at a speed of 160 mph. The five phases were:-

1. Receiver flight plan signalled to tanker base before take off.
2. Position reports sent from receiver to tanker base, giving rendezvous, fuel required and ETA.
3. W/T communication established between aircraft at 300 miles.
4. VHF communication established at 70 to 100 miles.
5. Final interception made by Eureka/Rebecca radar.

During 1947, B.E.A.C. was still short of aircraft for its services, and various types were leased by the airline for operation on both its internal and international services, and in April 1947, four Airspeed Consuls belonging to Westminster Airways were operated on the London to Helsinki service of B.E.A.C. These flights started on April 23 1947, and were flown from Croydon to Helsinki via Amsterdam, Copenhagen, Gothenburg and Stockholm. Another independent airline in the news at this time was Hunting Air Travel who took delivery of their first Vickers Viking on May 5 1947. Operated as a twenty one seater, it was soon joined by a second of the type. Other independent airlines were beginning to find work difficult to find, and in May 1947, Chartair and British American Air Services merged in order to try to rationalise their operations. Subsequently, Chartair operated a fleet of Consul aircraft on passenger services from Croydon and Hanworth, while British American Air Services concentrated on freight work with Haltons. Problems also troubled the Herts and Essex Aero Club who lost ten aircraft in a hangar fire on Sunday June 22 1947 - only three aircraft being saved by club members before the heat became too intense for them to save any more.

But other aircraft were also being lost - in accidents which became all too frequent, and most were caused by some error on the part of the crew or groundstaff. Some of these accidents unfortunately proved fatal and included:-

1. On September 27 1946, a Dragon Rapide of Railway Air Services (G-AFFF) crashed into high ground in conditions of low cloud and poor visibility. The report on the accident suggested that G-AFFF was flying on a series of bearings being sent to another Dragon Rapide - G-ADAJ - in an emergency, resulting from an erroneous message from G-ADAJ to the ground controller.

2. On April 16 1947, a Percival Proctor V of Guernsey Air Charter crashed into the sea between seven and ten miles off the French Coast near St. Jouin-sur-Mer. Of those on board, the pilot and two of the three passengers were drowned. The Proctor was forced to ditch, and although not recovered, it was established that G-AHZX had had an engine failure, although no reason was found for this.

3. On May 23 1947, Hunting's new Vickers Viking (G-AHPJ) made a heavy landing at Croydon on its return from a flight to Verona, and the undercarriage collapsed. During the subsequent inquiry it was found that the aircraft's loading was at fault in many respects including:-
   i) the C of G was 1.2 inches aft of the limit upon landing, and
   ii) the maximum permissable all up weight was exceeded by 189 lbs on take off at Verona.

4. On June 14 1947, a Miles Aerovan 4 (G-AISG) crashed immediately after take off from Croydon. Mr Malcolm Kerr (Captain) was killed and the radio operator and passenger injured. The aircraft was carrying 604 racing pigeons from Ringway to Guernsey and was found to be 446 lbs above its maximum all up weight. During the inquiry, no evidence cound be found of pre-crash failure of the aircraft or its equipment, and the findings were that the accident was caused by an error of flying technique on the part of the pilot who attempted to climb at too steep an angle, resulting in a loss of height in a semi-stalled condition.

5. On June 27 1947, a Miles Aerovan 4 (G-AJOB) operated by the Ulster Aviation Company, crashed into the sea south east of the South Rock Lighthouse while en-route Woodley to Newtownards. This was the result of the failure of the Aerovan to maintain height after its port engine had seized - the reason for this being the breakdown of the port engine-driven oil pump.

6. On November 20 1947, an Avro Anson of British Air Transport (G-AIWW) crashed at St. Boniface Down, Ventnor Isle of Wight while en route Croydon to Jersey. The pilot and radio operator were killed, and it was established that 'WW was fifteen miles west of the normal track between Croydon and Jersey, and that in conditions of low cloud, the Anson flew into a radar pylon as a result of an error in navigation.

Of these six typical accidents of the time, only two are due to aircraft failure, and in both cases this was due to a failure of one power unit, which in the case of a single engined Proctor is not a happy situation. But the failure of one power unit in an Aerovan, can only help to stress the extremely poor performance of early post way multi-engined aeroplanes when power failure occurred. One of the problems facing pilots was the lack of navigational equipment available to them in the cockpit. In today's world of DECCA, LORAN, and INS, it seems incredible to believe that all many of these aircraft carried was one VHF communications box; a VOR or an ADF was a luxury only afforded by the state corporations or large aircraft flying long distances. And so, the only aids the average Anson charter pilot had was a compass and a VHF transceiver which enabled him to use VHF D/F where such a facility was available. These radios were often unreliable, and in many cases were not able to receive all the required frequencies, resulting sometimes in the flights having to be conducted 'non-radio'. Another problem was the inexperience of many of the crews at this time, and the combination of an inexperienced Captain and a poorly maintained aircraft could prove fatal. Many companies by the end of 1947, were however beginning to pride themselves on the high standard and efficiency of their operation. Airlines in this category included Transair who operated a fleet of Ansons and Consuls on charter services from Croydon carrying passengers, freight and newspapers. Their standard of service was so high that they are still remembered by many today, as a shining example of how to operate an airline with outdated aircraft, and to succeed.

With cheap aircraft still available, many companies were still springing up and pleasure flying still proved immensely popular with a nation of aviation minded people - Mr Hugh Kennedy, an ex-test pilot working for the Miles Aircraft Company, flew 10/- joy rides in his Miles Nighthawk every Saturday and Sunday between two o'clock and dusk at Woodley Aerodrome during 1947 in fine weather. At Northolt Aerodrome where a spectator's enclosure had been opened, and for a small extra charge visitors could have a conducted tour of the airfield, Congo Air Charter offered 17/6d and 28/- joy rides over London in Airspeed Consuls. These are just two examples of the many companies devoting almost their entire time to this type of activity.

On November 1 1947, the last scheduled service from Croydon was made at 16.00 hours by a Dragon Rapide

(G-AGIF) operating the last B.E.A.C. service to Guernsey. Subsequently, Croydon was closed for regular scheduled air services, although later this was to change when private companies restarted their scheduled services.

1948 started with the tragic loss of 'Star Tiger', a British American Airways Avro Tudor en route London to Bermuda. The aircraft disappeared on the night of January 29 without a trace. The Tudor had already had a very chequered career before this, being rejected by B.O.A.C. on its North Atlantic services, and later rejected by the Corporation altogether. On February 23, a Tudor 1 (G-AGRI) left the United Kingdom for Nairobi and Khartoum for trials, and returned to Boscombe Down on completion of the tests reporting that snags found on earlier proving flights to Africa had now been ironed out. On March 24 1948, the ban on the Tudor IV was lifted to permit the type to be used by British South American Airways for carrying freight only. However on April 23 1948, the Minister of Civil Aviation announced that the Avro Tudor IV would be restored to all British South American Airway's passenger services, except those services linking the Azores and Bermuda,until further flight trials had been completed. But further problems arose when 'Star Ariel' was lost en route Bermuda to Jamaica on January 17 1949. After this accident, the Tudor was retired and the airframes were either stored or scrapped. The Tudor's first commercial service had been operated on September 30 1947, and less than eighteen months later the type had been withdrawn completely.

During 1948, Airwork had found profitable work for their new fleet of Vikings,operating services on behalf of the British Government,on the East African Ground Nuts scheme. During February, Airwork flew 97,796 miles carrying 1,329 passengers and 1,467,742 lbs of freight on these services. Also in February 1948, B.O.A.C. moved out of its Bovingdon base, resulting in Hunting Air Travel moving into Bovingdon during the same month. During the spring of 1948 numerous British charter airlines applied to B.E.A.C., for permission to operate, as Associates of the Corporation, various ferry and feeder air routes within the United Kingdom. Both B.E.A.C. and the Ministry of Civil Aviation had to approve these services, and the agreements which were drawn up only lasted for a period of six months, which gave the companies very little security over these services. In addition, the charter companies were entirely responsible for the services, bearing any losses on the routes operated, although of course the airlines also accepted any profits made on the services. By the end of May 1948, B.E.A.C. and the Minister of Civil Aviation had authorised Cambrian Air Services and Western Airways to operate a scheduled passenger service between Cardiff and Weston-Super-Mare. Although this marked the first real step towards private companies operating scheduled services once more, in May 1947 Morton Air Services operated their Rapides on a temporary scheduled service between Croydon and Castle Bromwich during the British Industries Fair at Birmingham. Despite the ban on the operation of scheduled services by private airlines, many airlines had got round it by operating what were loosely termed 'tourist' services, and these were especially popular on services to Jersey and the Isle of Man. However, by the end of 1948, the airlines found it easier to enter into an agreement with B.E.A.C., whereby they would operate a scheduled service as an Associate of the state airline. Unfortunately, there were some airlines who failed to operate as such, and with no agreement whatsoever, sold tickets to passengers as if they were a scheduled airline. One such airline was Air Transport Charter who were later prosecuted for what amounted virtually to the operation of regular scheduled services between Blackbushe and Jersey. During 1948 nine independent airlines operated scheduled internal passenger services under Associate Agreements with B.E.A.C. The companies were as follows:-

Air Enterprises Ltd.	Southampton to Cowes. Southampton to Cowes to Sandown. Gatwick to Cowes.
Barclays International Airways Ltd.	Cowes to Hurn. Croydon to Cowes to Hurn.
Brooklands Aviation Ltd.	Shoreham to Southampton.
Cambrian Air Services Ltd.	Weston-Super-Mare to Cardiff.
Lancashire Aircraft Corporation Ltd.	Leeds to Ronaldsway. Blackpool to Southport.
Northern Air Charter Ltd.	Newcastle to Ronaldsway.
Patrick Aviation Ltd.	Birmingham to Ronaldsway.
Scottish Aviation Ltd.	Prestwick to Blackpool.
Western Airways Ltd.	Weston-Super-Mare to Cardiff.

During 1948 a total of 12,896 passengers was carried on these services at an average passenger load factor of 42.8 per cent. B.E.A.C.'s British division, during 1948, returned passenger figures of 367,763 passengers carried at a passenger load factor of 62 per cent.

In July 1948, the Russians blockaded Berlin resulting in the enforced exile for the residents of West Berlin living in the French, American and British sectors. With no access available by rail, road or water, there was only one way left to get much needed supplies such as food, clothes, fuel and other essentials of life, to the people of Berlin and that was by air. On July 27 1948, Flight Refuelling started flying fuel into Berlin from Buckeburg using three Lancaster tankers. The United States Air Force and the Royal Air Force also took part using a mixture of C-54s, Hastings and Yorks. August 4 marked the first day of the Civil airlift, and on this day the following airlines started flying sorties:-

From Wunstorf - Bond Air Services (one Halifax), and Scottish Airlines (one Liberator).

From Fassberg - Scottish Airlines (nine Dakotas), Air Contractors (three Dakotas), Trent Valley Aviation (one Dakota), Air Transport Charter (one Dakota), Kearsley Airways (one Dakota), and Westminster Airways (one Dakota).

From Finkenwerder - Aquila Airways (two Sunderlands).

Ciros Aviation joined the airlift with one Dakota on August 6 based at Fassberg, Eagle Aviation joined on August 26 with one Halifax, but all Scottish Airlines aircraft had been withdrawn by August 27. The operation of these sorties took place day and night, and reliability was essential to a smooth operation. Airflight flew two Tudors on the airlift, and the airline only possessed two night rated Captains. When the airline's chief pilot, Captain Utting, was knocked down and killed by a lorry while on the tarmac at Gatow on December 8, this left Airflight with just one night rated Captain - the airline's founder and Managing Director,Air Vice Marshall D.C.T. Bennett. With no other captains available, he was forced to carry out all the night flying himself, and many times the aircraft was not in a serviceable state, but to ground an aircraft during the airlift would have meant a serious loss of revenue. On one occasion one of the Airflight Tudors developed a glycol leak in one of the engines, resulting in it overheating, but if this had been discovered by the authorities, it would have meant enforced withdrawal of the aircraft from the airlift. So to overcome this, on start up all four engines were started, and shortly after take off, the unserviceable engine was shut down once the aircraft was out of sight of the airfield, and the aircraft continued its flight on three engines. Just one example of the way in which many companies operated during the airlift.

For many airlines, the airlift had provided a much needed source of income at a time when work for many aircraft was difficult to find. During September and October 1948, many more airlines joined the airlift including Skyflight (September 17), Silver City Airways (September 18), British Nederland Air Services (September 21), British South American Airways (September 23), Transworld Charter (September 23), Hornton Airways (September 24), World Air Freight (October 6), the Lancashire Aircraft Corporation (October 16), Sivewright Airways (October 19) and British Overseas Airways Corporation (October 20). B.O.A.C.'s part was small, operating three Dakotas. The Corporation withdrew on November 25, and the following day marked the last sortie flown by a Dakota. Between November 10 and 22 1948, as each airline's contract expired who flew Dakotas, the aircraft were withdrawn as part of a reorganisation of the civil supply lines. The new plan was to get the greatest weight of cargo into Berlin in the shortest time, which meant that standardisation on the larger types of aircraft was essential, and coupled with this was the fact that there was an acute Dakota spares shortage.

By the time the 5,000th landing had been made by a civil aircraft on December 16, some 26,170 tons of supplies had been carried, and January's liquid fuel target for the month was 145 short tons of diesel fuel and 75 tons of MT petrol daily. By April, this target had increased to 350 tons daily, and by May to 550 tons daily. But the operation had its problems - one of which was highlighted on April 30 1949 when one of World Air Freight's Halifaxes crashed north of Tegel in the Russian Zone with the loss of four crew. On May 10 1949, another aircraft touched Russian soil, when one of Flight Refuelling's Lancastrians (G-AKDP) force landed in the Russian Zone, fortunately with no casualities. But two days later, the Russians ended their blockade of Berlin, and on May 31 1949, the 100,000th ton of supplies was delivered to Berlin. On July 13, the liquid fuel target was lowered to 130 tons daily, and on August 16 1949, the airlift ended. During the civil airlift, a total of 144,500 tons of food, fuel and other essentials had been delivered by midnight on August 15 1949 on a total of 21,785 flights by civilian aircraft. 54,325 tons of food and 87,474 tons of liquid fuel were delivered, and the number of sorties operated by each airline was as follows:-

Flight Refuelling, (Lancaster tankers)	4,438
Lancashire Aircraft Corporation (Halifax freighters 183, tankers 2,577)	2,760
Skyways, (York freighters 1,081, tanker 13, Lancastrian tankers 1,655)	2,749
Bond Air Services, (Halifax freighters)	2,577
British South American Airways, (Tudor freighters 231, tankers 2,331)	2,562
Eagle Aviation, (Halifax freighters)	1,054
Airflight, (Tudor freighter 85, tankers 836, Lincoln freighter 1, tanker 45)	967
Westminster Airways, (Dakota freighter 228, Halifax freighters 176, tankers 368)	772
British American Air Services, (Halifax freighters 97, tankers 564)	661
World Air Freight, (Halifax freighters)	526
Scottish Airlines, (Dakota freighters 101, Liberator freighter 15, tanker 381)	497
Air Contractors, (Dakota freighters)	386
Ciros Aviation, (Dakota freighters)	328
Aquila Airways, (Sunderland freighters)	265
Kearsley Airways, (Dakota freighters)	246
Silver City Airways, (Bristol Freighters)	213
Air Transport Charter, (Dakota freighters)	205
Trent Valley Aviation, (Dakota freighters)	186
Transworld Charter, (Viking freighters)	118
Hornton Airways, (Dakota freighter)	108
British Overseas Airways Corporation, (Dakota freighters)	81
British Nederland Air Services, (Dakota freighter)	76
Airwork, (Bristol Freighters)	74
Skyflight, (Halifax freighters)	40
Sivewright Airways, (Dakota freighters)	32

The load carried by each aircraft into Berlin was as follows:- Dakota 3tons 2cwt, Halifax 6tons 10cwt, Tudor 9tons 10cwt, Lancastrian 5tons 5cwt, and Sunderland 9tons 0cwt.

The Berlin Airlift was one of Flight Refuellings' few commercial operations, and as no history of the company appears in the main text, here below is a fleet list of their aircraft during the airlift with the number of sorties flown, the hours flown and the tonnage carried by each aircraft:-

Registration	Sorties flown	Hours flown	Tonnage carried	Remarks
G-AGWI	226	597.14	1,381.0	c/n 1281
G-AGWL	361	881.25	2,422.6	c/n 1284
G-AHJU	438	1,175.59	2,429.1	ex. LM681
G-AHJW	40	130.20	221.0	ex. ED866, written off Andover 23.12.48.
G-AHVN	279	657.28	1,586.1	ex. PP744
G-AKDO	431	1,141.02	2,683.0	ex. KB729
G-AKDP	378	1,053.53	2,216.9	ex. FM185, written off Russian Zone 10 5.49.
G-AKDR	526	1,472.04	3,070.1	ex. FM186
G-AKDS	480	1,305.31	2,784.5	ex. FM187
G-AKFF	449	1,129.14	3,022.5	ex. TX284
G-AKFG	439	1,099.38	2,943.0	ex. TX286
G-AKTB	391	985.33	2,354.8	ex. VM378

All the above aircraft were Avro Lancastrians of varying marks apart from G-AHJU, G-AHJW and G-AHVN which were converted Avro 683 Lancasters.

During the airlift, Flight Refuelling carried a total of 6,975,021 gallons of liquid fuel with a total flight time of 11,310.40 hours. The mileage covered was 1,714,596 statute miles losing only two aircraft during the entire operation, and one of those was lost while returning to its United Kingdom base at Tarrant Rushton for routine maintenance. A remarkable performance for a company not normally engaged in fully commercial operations.

Once the airlift had ended, all the aircraft returned to the United Kingdom - many airlines sold their aircraft and closed down, while others such as the Lancashire Aircraft Corporation managed to find charter work for their Halifaxes. Of the forty one Halifaxes used on the airlift, nine crashed during the operation, and although thirty two returned to the United Kingdom, most suffered under the breaker's axe, and by the end of 1950 all had been withdrawn from use.

Shortly after the Berlin Airlift had started, another airlift began on August 31 1948 carrying fifty thousand gallons of milk from Nutts Corner, Belfast to Liverpool and Blackpool. The milk was flown across daily in ten gallon milk churns aboard a fleet of two Liberators, six Halifaxes and five Dakotas belonging to seven airlines. The Dakotas were able to carry up to fifty six churns, while the Halifaxes could carry up to seventy churns. This operation lasted until the end of October, and was operated in a similar way to a milk airlift arranged a year before when twelve thousand gallons were flown from Northern Ireland to England. Those airlines involved with the 1948 milk airlift were Scottish Airlines (two Liberators and two Dakotas), the Lancashire Aircraft Corporation (two Halifaxes), British American Air Services (two Halifaxes), World Air Freight (one Halifax) Air Transport Charter (two Dakotas), Ciro's Aviation (one Dakota) and Bond Air Services (one Halifax). In addition, a Halifax of VIP Services appeared on the airlift during October, but it is believed that this was on lease to one of the other airlines involved in the airlift.

With both airlifts over, some companies turned to other sources of income - including aircraft component manufacture, aircraft overhaul or other activities totally disconnected with aviation. Skyways, which had operated a large fleet of Lancastrians, Yorks, Doves, Dakotas and DC4s, was run down until only two Dakotas were available for ad-hoc charter work from Dunsfold. In addition, the company overhauled aircraft and also carried out the ferrying of aircraft - particularly ex Royal Air Force aircraft bought by foreign air arms. So the end of 1949, and the whole of 1950 were years of little progress for the independents.

However, Sir Winston Churchill and a new Government were returned in the 1951 General Election, and within two years the laws forbidding the operation of scheduled services by private airlines were repealed. Once again the independents were able to return to scheduled passenger work for their source of income - many airlines, though, realised that the operation of many of these scheduled services would require newer and better equipped aircraft than they were are present operating. But even the de Havilland Dove was out of many airline's shopping lists because of the high cost of the aircraft compared to its veteran counterpart - the Dragon Rapide. However, by 1954, the scheduled airlines included Air Kruise, Airwork, Aquila Airways, BKS Air Transport, Cambrian Air Services, Derby Aviation, Don Everall Aviation, Eagle Aviation, East Anglian Flying Services, Hunting-Clan Air Transport, Jersey Airlines, Manx Airlines, Morton Air Services, Olley Air Service, Scottish Airlines, Silver City Airways, Skyways, Starways, Transair and Wolverhampton Aviation. And so, within two years of the restrictions being lifted, the independents were operating a large network of scheduled services using aircraft varying in size from the Dragon Rapide to the Viking, Dakota snd Bristol Freighter.

During 1952, the first of the many Air Ministry trooping contracts was placed with an independent - Airwork. To operate these services, the company bought four Hermes from B.O.A.C., and these aircraft were flown on troopong runs to Kenya and the Egyptian Canal Zone. Operating as sixty eight seaters, the Hermes flew in Royal Air Force colours and with military serials from Blackbushe Airport. Many other airlines found these trooping contracts a welcome money spinner including Hunting-Clan, Skyways (who operated the Lancashire Aircraft Corporation fleet of Yorks in addition to their own), Britavia, Silver City Airways, Scottish Airlines, Transair and Air Charter. Air Charter had been the brainchild of Mr. Freddie Laker who realised that although the York was a satisfactory trooping aircraft in 1952, it would no longer be in the very near future. Accordingly, Mr. Laker and another of his companies - Aviation Traders of Southend - searched for a York replacement, and in September 1953 announced that Aviation Traders had purchased, from the Ministry of Civil Aviation, a fleet of Avro Tudors.

These aircraft consisted of four Mk.1s, two Mk.3s and four Mk.4s. Eighty eight Merlin engines were also bought, and virtual rebuilding of one of these aircraft (G-AGRI), then took place. This process included a) the removal of pressurising equipment; b) the introduction of new cabin heating and ventilation systems; c) the re-routing of electrical cables and re-positioning of electrical and radio equipment; d) the incorporation of an additional emergency exit; e) the fitting of later types of wheels and airscrews; and f) the addition of passenger amenities.

This 'new' Tudor was granted a full C of A on February 2 1954, and during the tests, it was found that the performance of this aircraft was markedly better than in the same tests conducted by the Ministry of Supply on the original aircraft in 1946. On Sunday February 7 1954, Air Charter hoped to operate its first Tudor passenger service from Stansted to Hamburg, but due to ice on the Stansted runways, this flight had to be postponed. But, one week later, on February 14, Avro Tudor 1 - G-AGRG - made the company's first proving flight between Stansted and Hamburg and return. Fitted with forty two passenger seats (thirty eight rearward and four forward facing), the aircraft reached Hamburg in 2 hours 54 minutes captained by Captain E.N. Jennings - Air Charter's chief pilot. Hamburg was one of Air Charter's centres of operations with, at this time, two Yorks and two Bristol Freighters based there operating freight flights along the Berlin Corridor. The fleet of Tudors remained in service with Air Charter for just over five years giving sterling service on routes to the Middle and Far East on trooping runs, and in addition on a colonial coach service from Stansted to Idris and Lagos. By the end of 1959, all the Tudors had been retired from service, being replaced by the more modern Britannia.

By 1954, the charter companies had also started to operate inclusive tour charters, and for many airlines their whole growth would be based on the expansion of this market. But in 1954, the types of aircraft operated on these services rarely had a cruising speed in excess of 170 knots resulting in trips to Palma taking six hours from London, and some operators such as Starways made refuelling stops en route at Bournemouth and Jersey. Favourite destinations at this time included Perpignan, Pisa, Palma, Tarbes, Biarritz and Balbao, and inclusive tours to North Africa were certainly too far for the majority of the aircraft types then in operation by the independents. 1954 and 1955 saw the steady expansion of many airlines' scheduled services especially those of Hunting-Clan who were trying to develop a large network of scheduled passenger flights from Newcastle to many destinations both in the United Kingdom and Europe. During 1955, the company operated Viscounts on the London to Newcastle run in its attempts to boost these services from Newcastle. However, after several years of developing these services, the airline was forced to abandon them due to continuing losses, and on November 1 1955, the routes passed to Dragon Airways.

And so ten years had passed since the end of the Second World War - a turbulent time for most British independents, and it was not until the mid '50s that the private sector of this industry really settled down to develop into a major industry of the 60s and 70s. Apart from the setback of February 1947, the industry had been temporarily boosted by the Berlin Airlift, but the depression following the airlift caused many airlines to close. The reason for this was mainly due to the lack of work on the home market, and the inability to tender successfully for contracts abroad due to the poor standard of aircraft operated by the majority of British airlines. In 1950, the independents could only muster a couple of Yorks and one Tudor as long range charter aircraft. However, the next ten years saw sustained growth in the industry, but 1961 proved a particularly bad year, with many British independents closing down including Overseas Aviation, Air Safaris and several smaller carriers. 1960 had seen the formation of British United Airways who, under the management of Mr Freddie Laker, were to become the major British independent flying a large fleet of BAC 1-11s and Vickers VC 10s on scheduled passenger services,all over the world.

1968 was another bad year for the independents. At the time, the Government had imposed a £50 limit on the amount of money that one could spend abroad on a foreign holiday, causing many holidaymakers to stay in the UK for their holidays. This in turn meant a drop in the number of people flying abroad, and at the end of the year the collapse of British Eagle and Transglobe shook the industry in a similar way to the collapse of Courtline and Donaldson in August 1974.

A C E

Aviation Charter Enterprises Ltd.
ACE Freighters Ltd.
Ace Scotland Ltd.

Bases: Main operations base was Gatwick Airport, although the DC4 maintenance was carried out at Lasham, and L749 at Coventry.

On March 1 1964, Europe's first exclusively all-cargo airline began operations, when Aviation Charter Enterprises flew its first charter from Gatwick with its Constellation. The company was formed in September 1962 by four people - Captain L.C.Chegwidden, a New Zealander who acted as managing director; Mr.A.Coulson who became commercial director; Mr.A.McGill, technical director; and Mr.W.Kellard, operations manager. The airline applied to the A.T.L.B. for an 'E' licence for its operations, and this was granted early in January 1964. The airline initially catered for medium to long haul requirements with its Constellation - this aircraft had a large, rear freight door and with its 26,000 lb capacity proved to be ideal for the airline's operations.

Trading as ACE Freighters, the Constellation flew ad-hoc freight charters from many United Kingdom airports to numerous destinations abroad, particularly in the Middle and Far East. During the year two Douglas DC4 freighters were acquired and the company also tendered successfully for several Ministry of Defence freight contracts, and subsequently many of these flights were operated from Lyneham and Abingdon to the Middle and Far East. In November 1964, ACE bought the entire South African Airways Constellation fleet, and the first of these aircraft arrived at Gatwick in December 1964.

Eventually all four aircraft arrived at Coventry Airport where the company had leased hangar space, and had set up maintenance facilities. During 1965 one of these Constellations was put into service, but the remaining three aircraft never gained certificates of airworthiness, and were stripped for spares. During the year, two Euravia Constellations were purchased and a third leased. The Ministry of Defence charters continued throughout 1965 and into 1966, and these provided the airline with most of its revenue. The Constellations regularly flew freight from Lyneham and Abingdon to the Middle and Far East, to destinations including Aden, Malta, Cyprus, El Adem and Singapore. It was while one of the Constellations (G-ANTF) was operating one of these services that it carried out a perfect wheels-up landing at Aden on March 8 1966. The Captain managed to feather all four propellers and position them to avoid damage before touch down, and although fire did break out in the nose wheel area shortly after touch down, it was quickly extinguished, and after minor repairs the aircraft returned to England. The two DC4s were maintained by Dan-Air at Lasham, but were withdrawn from service in the spring of 1966, and the company standardised on an all-Constellation fleet. The DC4s were both used for a series of flights between Southend and Antwerp during 1965 on behalf of the Ford Motor Company. During the 1966 dock strike, ACE found ready work flying various stranded goods into and out of this country. The Constellations flew numerous charters carrying goods such as cigarettes out of many airports including Gatwick, Southend, Liverpool, Glasgow and Newcastle to European capitals such as Paris and Amsterdam.

In July 1966, a subsidiary company was formed named Ace Scotland to operate inclusive tour charters from Glasgow to many Mediterranean resorts during the 1966 summer season. One Constellation (G-ASYF) was transferred to this company for this purpose, and after conversion and painting in a new pale blue and white livery, the aircraft positioned to Glasgow. This marked ACE's one and only entry into passenger carrying, because on September 14 1966 all flying ceased, and the company was put into liquidation. This date also marks the end of twenty years of British Constellation operations, and many of ACE's aircraft were never to see commercial service again, indeed many never even left their graveyard at Coventry.

Fleet:

Douglas DC4	G-APEZ	c/n	42921	dd. 8.64,	wfu. 7.66.
Douglas C54A Skymaster	G-ASEN	c/n	10412	dd.11.64,	sold. 4.66.
Lockheed L749A Constellation	G-ALAK	c/n	2548	dd. 8.65,	ISOC.
	G-ALAL	c/n	2549	dd. 8.65,	ISOC.
	G-ANTF	c/n	2504	dd. 2.64,	ISOC.
	G-ANUR	c/n	2565	lsd. 8.65,	ret. 9.66, leased from Euravia.
	G-ASYF	c/n	2630	dd.12.64,	ISOC, Ace Scotland Aircraft.
	G-ASYS	c/n	2623	dd. 1.65,	never entered service.
	G-ASYT	c/n	2631	dd. 2.65,	never entered service.
	G-ASYU	c/n	2632	dd. .65,	never entered service.

ADIE AVIATION

Adie Aviation Ltd.

Base: Croydon Airport.

Adie Aviation was a small air charter and aircraft maintenance company formed early in 1948 at Croydon Airport. Initially the company solely dealt in the sale of second hand aircraft, and in their maintenance and overhaul. The company became recognised as Rapide specialists, and many former Royal Air Force Dominies were converted by the Company for use by the many post-war civilian operators. In addition to the normal Rapide conversions, Adie Aviation also fitted extra large cabin doors, long range fuel tanks, photographic windows and many different types of radio equipment. Throughout its association with the Rapide, Adie Aviation retained one aircraft for use on air taxi and light charter work - their aircraft were often used by the Press, and on day-trips to racemeetings in the British Isles. However, the company still concentrated on aircraft repair and conversion at Croydon, and the company hangered and maintained other operator's aircraft at Croydon Airport. The last Rapide was sold in April 1954 after five years of charter work with this type of aircraft - Adie was now in the process of converting numerous Royal Air Force Tiger Moths, for sale to civilian customers both at home and abroad, and although a Miles Mercury was retained as a company hack, all charter work had ceased by 1956.

Fleet:

DH.89A Dragon Rapide	G-AFMF	c/n	6432	dd. 7.49,	sold 4.50.
	G-AGJG	c/n	6517	dd. 7.50,	sold 4.54.
	G-AKZW	c/n	6896	dd. 6.49,	sold 7.50.
Percival Proctor 1	G-AIEE	c/n	K.280	rd. 4.48,	sold 5.48.
Miles M.28 Mercury 6	G-AHAA	c/n	6268	rd. 1.52,	sold 9.56.
Airspeed AS.65 Consul	G-AJGG	c/n	5123	dd. 9.52,	sold 3.53.

AER ARANN

Aer Arann Teo.
United Developments (Galway) Ltd.

Bases: Oranmore, later Kiltulla and later Galway Airfields, Eire.

Aer Arann was formed in 1970 as a charter and scheduled third level operator within the Republic of Ireland The airline was formed by a group of local businessmen, and the airline's chief executive, managing director and chief pilot is Captain Bill Wallace, who was formerly a Captain with Aer Lingus. Services started in July 1970 with one Britten-Norman Islander which was used on scheduled passenger routes linking Shannon with two of the larger Aran islands, Inishmore and Inishman. Initially, the company operated from its own airfield at Oranmore, but later the airline moved its base to Kiltulla. Kiltulla soon became too small for Aer Arann's expanding operations, and Captain Wallace began to look around for a new base. Fortunately, a German firm - the Steiner

Company - was building itself a new airstrip and hanger beside its factory in Galway. This tarmac airstrip is now rented to Aer Arann, and the airfield is now the airline's base of operations.

Today, in addition to their scheduled services, Aer Arann flies a wide variety of charter services with its fleet of Islanders. The airline flies regularly from Dublin to the Channel Islands, and on September 7 1974 a thrice weekly charter service was inaugurated from Letterkenny, Donegal to Glasgow. Early in 1975, a weekly Tuesday or Wednesday charter flight carrying newly printed money from Newcastle to Dublin was also started, but in October 1975 two of the company's Islanders were damaged in accidents in Ireland, and Aer Arann's services were temporarily curtailed until the aircraft were repaired. Aer Arann's Islanders have also become familiar sights at many other English airports on charter flights, but Aer Arann's prime task still lies in the operation of its scheduled third level services to the Aran islands with its fleet of Islanders.

Fleet:

BN.2A Islander	EI-AUL	c/n	180	dd. 7.70, sold 5.73.
	EI-AWM	c/n	653	dd. 5.73.
	EI-AYN	c/n	704	dd. 4.74.
	EI-BBA	c/n	444	dd. 6.75.

AERIAL ENTERPRISES

Aerial Enterprises Ltd. Base: Booker Aerodrome.

This charter operator commenced operations from Booker in March 1968. Two Dragon Rapides were operated on general charter work, but after little more than one year's operations, the company closed down. The Rapides were often used for parachuting and aerial photography from Booker.

Fleet:

DH.89A Dragon Rapide	G-AGJG	c/n	6517	dd. 3.68, ISOC.
	G-AJGS	c/n	W1001	dd. 3.68, ISOC.

AERIAL TRANSIT CO

Aerial Transit Company Ltd. Base: Croydon Airport.

The Aerial Transit Company came into existence on January 10 1946 when this company was registered as an airline company. The founders were Lord Grimthorpe and Frederic C.L.Jaques, who were both directors of Martin Hearn Ltd., who specialised in aircraft overhaul and maintenance at Hooton Park in Cheshire. The Aerial Transit Company set up its base at Croydon Airport and was formed to establish, maintain and work air services in any part of the world. In November 1946 numerous Ansons were purchased from the Royal Air Force with the intention of converting these aircraft for passenger and freight operations. The aircraft were delivered direct to Martin Hearn at Hooton Park for conversion. However, many never gained Certificates of Airworthiness, and others were sold to other operators immediately after conversion. At the most, only two aircraft entered service with the Aerial Transit Company, but all operations had ceased by May 1949.

Fleet:

Avro 652A Anson 1	G-AHXT			dd. 2.47, sold 2.47, previous identity not known, never entered service.
	G-AIRL	ex.	MG970	dd.11.46, never entered service.
	G-AIRM	ex.	MG585	dd.11.46, sold , C of A 5.48.
	G-AIRN	ex.	NK667	dd.11.46, sold 5.49, C of A 5.48.
	G-AIRO	ex.	N9895	dd.11.46, never entered service.
	G-AIRW	ex.	MH167	dd.11.46, sold , C of A 10.48.
	G-AIRX	ex.	AX232	dd.11.46, sold 4.47, never entered service.
	G-AHRD	ex.	EG633	dd. 7.46, sold 4.47.
	G-AITL	ex.	EG324	dd.11.46, sold 7.47, never entered service.
	G-AIZH	ex.	MG472	dd.12.46, never entered service.

AER LINGUS

Aer Lingus Teo.
Irish International Airlines.
Aerlinte Eireann.

Bases: Initial base was Baldonnel and current base is Dublin Airport.

Aer Lingus Teoranta's history can be traced back to May 22 1936 when, on that date, a private company of that name was registered as an airline. The company was backed by a British airline - Blackpool and West Coast Air Services, and on May 27 operations started with a five seat de Havilland Dragon EI-ABI. The first service was a flight from Baldonnel, Aer Lingus' base, to Bristol's Whitchurch Airport, and this service was flown daily in conjunction with Blackpool and West Coast Air Services under the title of Irish Sea Airways. On May 30 a service was opened linking Baldonnel with the Isle of Man on a thrice weekly basis. Further expansion came in September with the delivery of a Dragon Express, which on September 14 inaugurated a daily Dublin to Liverpool scheduled service. Also during the month, the Dublin to Bristol service was extended to Croydon.

On April 5 1937, the Irish Government backed Aer Rianta was founded to established international air services from Dublin. During 1937 and 1938,little expansion took place although the Dragon was replaced by a Dragon Rapide, and a second Dragon Express added during 1938. However, with the outbreak of war, all operations were temporarily suspended, although on October 28 1939 the Dublin to Liverpool service was reopened. During the war, this service continued, operated by a Dakota and two Dragon Expresses. Once the war had ended, Aer Lingus was able to settle down and to build up its network of scheduled services, reopening the Dublin to Croydon service on November 9 1945, and on June 17 1946 a Dublin to Paris service was started. The first Dublin to London service was flown by Dakota EI-ACA, and the Dublin to Shannon route was reopened on April 8 1946. From July 1 1946, the Dublin to Liverpool service was flown five times daily by Dragon Expresses, and the service to London was flown four times daily by Dakotas. Then in 1947 with the introduction of a fleet of Vikings, further expansion took place with routes being opened from Dublin to Manchester, Amsterdam, Paris, Belfast, Glasgow, Brussels and Rome; and from Shannon to Paris and London. On February 26 1947, Aerlinte Eireann Teoranta was formed as a subsidiary of Aer Rianta to operate transatlantic services, and on September 17 three new Constellations arrived at Shannon for use on these services. However, these services were indefinately postponed, and rather than let these new aircraft lie idle, they were introduced onto the services to Rome and London, and then in 1948 the aircraft were sold to B.O.A.C.

The Viking had not proved particularly popular with Aer Lingus, and with the cutback in certain services, the fleet was sold in 1948. The frequency of the Dublin to London service was increased to ten daily with the return of the Dakotas to this service, and in 1949 Aer Lingus opened new services from Dublin to Birmingham and Jersey. Over the next few years, gradual expansion of the airline took place, and with the expected increase in freight traffic, Aer Lingus ordered a fleet of Bristol Freighters in 1952, but the build-up in freight was not as rapid as Aer Lingus had hoped, and the aircraft were later sold. In 1952, new services were inaugurated from Dublin to Cardiff and Edinburgh. The Cardiff service started on June 13, and the Edinburgh service was opened on April 22 1952 - both services being operated by Dakota EI-ACE. In November the airline concluded an agreement with Seaboard and Western Airlines for operating the American airline's DC4s on transatlantic services from Dublin to New York, but permission was not granted by the American C.A.B., and the plans were dropped.

In November 1951, Aer Lingus had become the third airline in the world to order the Vickers Viscount, and in March 1954 the company took delivery of its first aircraft, later entering service with the airline on April 11 1954. Later, further Viscounts were ordered, and in June 1956 a fleet of Fokker Friendships was also ordered. Then, on September 1 1956, an agreement was signed with B.E.A. whereby B.E.A. was given traffic rights to Dublin as from April 1 1957. Thus, Aer Lingus now had competition on its routes to London, Manchester, Birmingham and Newcastle, but during 1957 new services were opened to Brussels, Dusseldorf, Frankfurt and Zurich. On April 28 1958, Constellation N1009C opened a thrice weekly service from Shannon to New York. The aircraft was leased from Seaboard and Western, and although flown by Seaboard crews, the cabin staff was provided by Aerlinte Eireann. Further Constellations were later leased when the service to New York became daily from June 1, although during the winter it returned to a thrice weekly frequency. With the success of this service, Aerlinte placed an order with Boeings for three B720-048s on March 11 1959, with delivery taking place late in 1960.

The first Friendships were delivered to Aer Lingus on November 19 1958, and these aircraft entered service on December 15 on the routes to Glasgow and Liverpool. On April 14 1959, the Dublin - Manchester - Dusseldorf service was extended to Copenhagen, and a route from Dublin to Blackpool was started on May 15. From January 1 1960, Aer Lingus and Aerlinte began to operate as one airline under the title of 'Aer Lingus - Irish International Airlines'. On June 3 1960, a twice weekly Dakota service was opened from Dublin to Cherbourg, and on October 25 the first B720 was handed over to Aer Lingus at Seattle, arriving at Shannon on November 18. After a series of route proving and crew training flights, the aircraft operated Aer Lingus' first scheduled jet service from Dublin, via Shannon, to New York on December 14. The Constellations were now returned to Seaboard, and all three B720s were in service by the end of April 1961. During 1962, services were opened from the new airport at Cork to Dublin, Bristol, Cardiff, London and Paris. Then, in August 1962, Aer Lingus announced plans to open a network of vehicle ferry services from Dublin to Bristol and Liverpool, and from Cork to Bristol. For these services, Aer Lingus ordered two Carvairs from Aviation Traders, and the first aircraft arrived in Dublin on March 14 1963. On May 8, the Dublin to Liverpool service was started, followed by the Dublin to Bristol service on May 11, and the Cork to Bristol service on May 12. In mid-June, a Dublin to Cherbourg service was started, and with a good first season, a third Carvair was ordered for delivery in 1964. In addition to the vehicle ferry service, the Carvairs were used for all-freight operations during the night, and also during the winter months, when the car-ferry services were not flown. During the first season of operations, 4,100 cars and 11,000 passengers were carried by the Carvairs.

During the next two years, the B720s were replaced by B707s on the trans-Atlantic services, and BAC 1-11s were introduced on many of the European services from 1965 onwards. The first 1-11 service was flown on June 6 1965 on the Dublin - Cork - Paris run. Also in June, new services were opened from Cork and Shannon to Manchester. One of Aer Lingus' first 1-11 charter flights was made in January 1966 when one of the aircraft carried the Glasgow Celtic football team from Prestwick to Tbilisi in Russia on January 22. The aircraft flew via Copenhagen and Moscow and carried an Aer Lingus crew who were assisted by two Soviet navigators. The Friendships were phased out early in 1966, and replaced with a fleet of K.L.M. Viscounts, and on March 9 1966, Aer Lingus placed an order for two Boeing 737s for use on the Dublin to London service. On May 9 1966, a twice weekly service was opened from Dublin to Montreal and Chicago, and in January 1967 a new service was started from Dublin to Brussels and Munich. A Dublin to Madrid service was opened on May 29, but the car-ferry services were now proving to be uneconomical, and the Carvairs were withdrawn from service and advertised for sale. The order for B737s was increased to eight aircraft during 1967, and on April 2 the first aircraft arrived at Dublin entering service later in the month on the schedules to Paris and London from Dublin. Gradually as the B737s entered service, the Viscounts were withdrawn and either broken up or leased out to other operators including British Air Ferries and the Swiss airline S.A.T.A. of Geneva. Two Boeing 747s had been ordered in January 1967, and the first of these aircraft arrived in March 1971, and although initially used on trans-Atlantic services from Dublin and Shannon, loads did not prove sufficiently high for both aircraft to be used on these services, and susequently one aircraft has been leased to Air Siam. In addition, of the B737s and B707s have been leased out to other operators including Transair of Canada, Air Algerie and Zambia Airways.

Today, Aer Lingus is an all-jet operator, and some of the less economical routes, which would not support jet aircraft, have been withdrawn including the Cork to Bristol, and the Dublin to Blackpool services. However, Aer Lingus does operate a large number of inclusive tour charters, particularly at weekends during the summer months, from Dublin and Cork to most of the popular Mediterranean resorts, and during the early part of the summer many flights are operated to Tarbes in the south of France, carrying pilgrims to nearby Lourdes. In recent years, the Boeing 707s have been used on trans-Atlantic passenger charters to many of the smaller British airports including Bristol, Liverpool and Newcastle.

Fleet:

DH84 Dragon	EI-ABI	c/n 6076	dd. 5.36,	sold 1.38,	'Iolar'.
DH86B	EI-ABK	c/n 2338	dd. 9.36,	sold 10.46,	'Eire'.
	EI-ABT	c/n 2336	dd.10.38,	sold 11.46,	'Sasana'.
DH89 Dragon Rapide	EI-ABP	c/n 6341	dd. 2.38,	sold 2.40,	'Iolar ll'.
Lockheed 14	EI-ABV	c/n 1497	dd. 6.39,	sold 5.40.	
	EI-ABW	c/n 1498	dd. 6.39,	sold 5.40.	
Lockheed Hudson	EI-ACB	c/n 1812	dd. 8.45,	sold 1.47.	
Supermarine Walrus	EI-ACC	c/nCS21840	rd. 8.45,	sold 11.46,	never entered service.
Miles M.65 Gemini	EI-ACW	c/n 6320	dd. 6.47,	sold 6.48.	
	EI-ADM	c/n 6520	dd.12.47,	sold 6.49.	
Airspeed AS.65 Consul	EI-ADB	c/n 5134	dd. 4.48,	sold 6.49.	
	EI-ADC	c/n 5151	dd. 4.48,	sold 6.49.	
Douglas DC3 Dakota	EI-ACA	c/n 2178	dd. 4.40,	written off	Shannon 18 June 1946.
Douglas C47 Dakota	EI-ACD	c/n 9140	dd.12.45,	sold 8.63,	'St.Patrick' later 'St.Senan'.
Douglas DC3 Dakota	EI-ACE	c/n 42956	dd. 2.46,	sold 7.64,	'St.Colmoille' later 'St.Celsus'.
	EI-ACF	c/n 42957	dd. 2.46,	written off	Spernall Heath 1.1.53, 'St.Kievan'
Douglas C47 Dakota	EI-ACG	c/n 4579	dd.12.45,	sold 4.59,	'St.Malachy'.
Douglas C47A Dakota	EI-ACH	c/n 12893	dd. 1.46,	sold 5.59,	'St.Bridgid' later 'St.Mel'.
Douglas C47 Dakota	EI-ACI	c/n 9036	dd.12.45,	sold 1.61,	'St.Aidan', later 'St.Ceannac'.
Douglas C47A Dakota	EI-ACK	c/n 19503	dd. 1.46,	sold 3.59,	'St.Albert'.
	EI-ACL	c/n 11861	dd. 4.46,	sold 5.58,	'St.Declan'.
	EI-ACM	c/n 12899	dd. 1.46,	sold 5.58,	'St.Fintan'.
	EI-ACT	c/n 12471	dd.10.46,	sold 5.59,	'St.Colman'.
Douglas C47B Dakota	EI-ADW	c/n 26107	lsd. 5.48,	ret. 10.48.	
	EI-ADX	c/n 26728	lsd. 5.48,	ret. 10.48.	
	EI-ADY	c/n 26997	lsd. 5.48,	ret. 11.48.	
	EI-AFA	c/n 32724	dd.11.48,	sold 10.63,	'St.Lawrence O'Toole' later 'St.Kevin' later 'St.Fergal' later 'St.Declan'.
Douglas C47A Dakota	EI-AFB	c/n 20453	dd.11.48,	sold 1.58,	'St.Brendan' later 'St.Cillian' later 'St.Jarlath'.
	EI-AFC	c/n 20135	dd.11.48,	sold 6.64,	'St.Enda'.
Douglas C47B Dakota	EI-AFL	c/n 33447	dd.11.48,	written off	Snowdonia 10.1.52, 'St.Kevin'.
	EI-AHG	c/n 32761	dd. 3.55,	sold 2.64,	'St.Ronan'.

ACE Freighters' first Constellation placidly sits on Southend's apron on August 11, 1964.

Transportraits/John Goring

Aviation Charter Enterprises' one and only entry into the passenger charter business came with the introduction of this Constellation G-ASYF in Ace Scotland's colours.

Guy Fawkes morning 1975 at Norwich, and Air Anglia's Herald G-ASVO awaits passengers for AQ202 to Aberdeen.

Tony Merton Jones

EI-AWM Aer Arann Islander — **Austin J Brown**

EI-AYN Aer Arann Islander — Ian MacFarlane

EI-ASE Aer Lingus Boeing 737 — **Tony Merton Jones**

EI-ALC Aer Lingus Boeing 720 — B N Stainer/A.P.N.

EI-AML Aer Turas Dragon Rapide — Ian MacFarlane

EI-AOR Aer Turas Douglas DC-4 — Ian MacFarlane

Aer Lingus' first Viking EI-ADF was delivered to Dublin in June 1947, but only flew with the airline until early in 1948, when this aircraft was sold.

Vickers Ltd.

Seen on final approach at Heathrow Airport, Aer Lingus' third BAC 1-11 EI-ANG has now been flying domestic and international services without an accident for over ten years.

British Aircraft Corporation

G-AIVF, pictured here at Speke Airport, Liverpool, was one of several Vikings used by Air Ferry on ad hoc and inclusive tour charter work.

P H Butler

Lockheed L749A Constellation	EI-ACR	c/n	2548	dd. 9.47,	sold 6.48,	'St.Brendan'.
	EI-ACS	c/n	2549	dd. 9.47,	sold 6.48,	'St.Patrick'.
	EI-ADA	c/n	2554	dd. 9.47,	sold 6.48,	'St.Bridget'.
	EI-ADD	c/n	2555	dd. 9.47,	sold 6.48,	'St.Kevin'.
	EI-ADE	c/n	2556	dd.10.47,	sold 6.48,	'St.Finbar' later 'St.Enda'.
Vickers 634 Viking	EI-ADF	c/n	208	dd. 6.47,	sold 2.48,	'St.Ronan'.
	EI-ADG	c/n	209	dd. 6.47,	sold 11.48,	'St.Senan'.
	EI-ADH	c/n	210	dd. 7.47,	sold 11.48,	'St.Celsus'.
	EI-ADI	c/n	211	dd. 7.47,	sold 2.48,	'St.Mel'.
	EI-ADJ	c/n	212	dd. 7.47,	sold 12.48,	'St.Flannan'.
	EI-ADK	c/n	213	dd. 9.47,	sold 8.48,	'St.Jarlath'.
	EI-ADL	c/n	214	dd. 9.47,	sold 8.48,	'St.Felim'.
Vickers 498 Viking	G-AGRU	c/n	112	lsd.12.46,	ret. 2.47,	leased from B.E.A.
Bristol 170 Freighter Mk.31	EI-AFP	c/n	12827	lsd. 3.52,	ret. 10.52,	leased from Bristol Aircraft Co.
	EI-AFQ	c/n	12937	dd. 6.52,	sold 10.55,	'St.Finbar'.
	EI-AFR	c/n	13072	dd. 7.52,	sold 11.55,	'St.Ronan'.
	EI-AFS	c/n	13074	dd.12.52,	sold 1.57,	'St.Senan'.
	EI-AFT	c/n	13076	dd. 1.53,	sold 10.56,	'St.Flannan'.
Vickers 707 Viscount	EI-AFV	c/n	030	dd. 3.54,	sold 5.60,	'St.Padraig' later 'St.Garan' later 'St.Colmcille'
	EI-AFW	c/n	031	dd. 3.54,	sold 1.60,	'St.Brighid' later 'St.Jarlath' later 'St.Melachy'.
	EI-AFY	c/n	032	dd. 3.54,	sold 12.59,	'St.Brendan'.
	EI-AGI	c/n	034	dd. 4.54,	sold 4.60,	'St.Lorcan O'Tuathail'.
Vickers 745 Viscount	EI-AJV	c/n	228	lsd. 3.58,	ret. 6.58,	leased from Vickers.
	EI-AJW	c/n	225	lsd. 4.58,	ret. 6.58,	leased from Vickers.
Vickers 803 Viscount	EI-AOE	c/n	177	dd. 2.66,	wfu. 11.69,	'St.Damhnait'.
	EI-AOF	c/n	176	dd. 2.66,	written off	Ashbourne 22.6.67, 'St.Finghin'.
	EI-AOG	c/n	172	dd. 5.66,	wfu. 10.69,	'St.Finian'.
	EI-AOH	c/n	180	dd. 6.66,	wfu. 11.70,	'St.Fiacre'.
	EI-AOI	c/n	179	dd. 7.66,	wfu. 10.71,	'St.Fergal'.
	EI-AOJ	c/n	173	dd.12.66,	wfu. 11.70,	'St.Flannan'.
	EI-AOL	c/n	175	dd. 2.67,	sold 3.73,	'St.Fintan'.
	EI-AOM	c/n	176	dd. 2.67,	written off	Irish Sea 24.3.68, 'St.Feidhlim'.
	EI-APD	c/n	174	dd.12.66,	wfu. 11.70,	'St.Finbar'.
Vickers 808 Viscount	EI-AJI	c/n	289	dd. 5.57,	wfu. 10.69,	'St.Gall'.
	EI-AJJ	c/n	290	dd. 1.58,	wfu. 12.69,	'St.Colomban'.
	EI-AJK	c/n	291	dd. 2.58,	sold 3.73,	'St.Cillian'.
	EI-AKJ	c/n	421	dd. 1.59,	sold 7.73,	'St.Colman', reregistered EI-AKO.
	EI-AKK	c/n	422	dd. 1.59,	written off	Lulsgate 21.9.67, 'St.Aodhan'.
	EI-AKL	c/n	423	dd. 3.59,	sold 3.73,	'St.Colmcille'.
	EI-ALG	c/n	312	dd. 2.60,	wfu. 11.69,	'St.Kieran'.
	EI-AMA	c/n	258	dd. 4.62,	wfu. 1.70,	'St.Conleth'.
Lockheed L1049G S.Constellation	N611C	c/n	4602	lsd. 5.58,	ret. 11.58,	'St.Brendan', leased from Seaboard
Lockheed L1049E S.Constellation	N1005C	c/n	4557	lsd. 5.58,	ret. 12.60,	'St.Brighid', leased from Seaboard
Lockheed L1049H S.Constellation	N1008C	c/n	4806	lsd.11.58,	ret. 5.60,	'St.Brendan', leased from Seaboard
	N1009C	c/n	4807	lsd. 4.58,	ret. 12.60,	'St.Patrick', leased from Seaboard
Fokker F.27.100 Friendship	EI-AKA	c/n	10105	dd.11.58,	sold 6.66,	'St.Fionntar'.
	EI-AKB	c/n	10106	dd.11.58,	sold 1.66,	'St.Fearghal'.
	EI-AKC	c/n	10107	dd.12.58,	sold 3.66,	'St.Fionnbharr'.
	EI-AKD	c/n	10109	dd. 1.59,	sold 1.66,	'St.Flannan'.
	EI-AKE	c/n	10110	dd. 2.59,	sold 6.66,	'St.Feidhlim'.
	EI-AKF	c/n	10118	dd. 5.59,	sold 1.66,	'St.Finghin'.
	EI-AKG	c/n	10119	dd. 5.59,	sold 1.66,	'St.Fiachra'.
Boeing 720-048	EI-ALA	c/n	18041	dd.11.60,	sold 10.72,	'St.Padaraig' later 'St.Pappin'.
	EI-ALB	c/n	18042	dd. 1.61,	sold 9.64,	'St.Brighid'.
	EI-ALC	c/n	18043	dd. 4.61,	sold 8.72,	'St.Brendan'.
Avaition Traders ATL.98	EI-AMP	c/n	06	dd. 3.63,	sold 5.68,	'St.Albert'.
	EI-AMR	c/n	08	dd. 4.63,	sold 2.68,	'St.Jarlath'.
	EI-ANJ	c/n	14	dd. 4.64,	sold 2.68,	'St.Senan'.
BAC 1-11 208	EI-ANE	c/n	045	dd. 5.65		'St.Mel'.
	EI-ANF	c/n	050	dd. 6.65		'St.Malachy'.
	EI-ANG	c/n	051	dd. 7.65		'St.Declan'.
	EI-ANH	c/n	052	dd. 9.65		'St.Ronan'.
Boeing 707-348C	EI-AMW	c/n	18737	dd. 6.64,	sold 6.72,	'St.Lawrence O'Toole'.
	EI-ANO	c/n	18880	dd. 4.65		'St.Brighid'.
	EI-ANV	c/n	19001	dd. 2.66		'St.Enda'.
	EI-APG	c/n	19410	dd. 7.67		'St.Senan'.
Boeing 707-349C	EI-ASN	c/n	18976	dd. 4.69		'St.Eunan'.
	EI-ASO	c/n	19354	dd. 4.69		'St.Canice'.
Boeing 737-248	EI-ASA	c/n	19424	dd. 3.69		'St.Jarlath'.
	EI-ASB	c/n	19425	dd. 4.69		'St.Albert'.
	EI-ASE	c/n	20220	dd.11.69		'St.Fachtna'.
	EI-ASF	c/n	20221	dd. 1.70		'St.Nathy'.
	EI-ASG	c/n	20222	dd. 2.70		'St.Cormac'.
	EI-ASH	c/n	20223	dd. 4.70		'St.Eugene'.
Boeing 737-248QC	EI-ASC	c/n	20218	dd. 9.69		'St.Macartan'.
	EI-ASD	c/n	20219	dd.10.69		'St.Ide'.
Boeing 737-248C	EI-ASL	c/n	21011	dd. .75		'St.Killian'.
Boeing 737-222	EI-ASK	c/n	19947	lsd. 5.74	ret. 9.74,	'St.Lawrence O'Toole' leased from United.
	EI-ASK	c/n	19947	lsd. 5.75		'St.Lawrence O'Toole' leased from United.
Boeing 747-148	EI-ASI	c/n	19774	dd. 3.71		'St.Comcille'.
	EI-ASJ	c/n	19775	dd. 3.71		'St.Padraig'.
Boeing 707-351C	EI-ASM	c/n	19263	lsd. 7.75		leased from Zambia Airways.

AER TURAS

Aer Turas Teo. — Base: Dublin (Collinstown) Airport.

Aer Turas was formed early in 1962 to undertake ad-hoc passenger and freight charters from its base at Dublin Airport. Operations began in June 1962 with a Dragon Rapide, and most charters were confined to within the United Kingdom and Eire, and during its two years with Aer Turas the aircraft did visit Belfast, Liverpool and Newcastle on charter flights. However, in March 1964 the Rapide was replaced by a Dakota, and this aircraft was also used on a variety of ad-hoc passenger and freight charters. During the summer of 1964 the Dakota flew regularly from Dublin to Birmingham and Bristol, and visits were also made to Cambridge, Liverpool, Manchester and Newcastle on charter flights. During the height of the summer, the Dakota made up to five flights each week into Birmingham from Dublin, with the occasional flight being made to Cork. In July the Dakota flew a charter from Dublin to Brussels, but the aircraft was rarely seen on the Continent, and financial problems in September 1964 brought Aer Turas' operations to an end at the end of the month - one of the last charters was a flight from Dublin to Bristol and Birmingham, and return, on September 29 1964. During the winter Aer Turas reorganised its operations,and flying was resumed in June 1965 with one Douglas DC4, which was used on a variety of passenger and freight charters from Dublin Airport.

During 1966 the fleet was expanded with the acquisition of a second DC4 and two Bristol Freighters. The DC4s were often used by Cambrian Airways for use on their scheduled and charter flights from Liverpool and Cardiff after the loss of one of their Viscounts at Liverpool. The Bristol Freighters were solely used for freight charter work from Dublin, and they were often used for the carriage of racehoses. In 1967 the Freighters were used to operate freight services on behalf of Aer Lingus from Dublin to Birmingham, Bristol, Heathrow, Liverpool and Manchester, and although these services continued into 1968, they were discontinued at the end of the year. Early in 1969 the company purchased the K.L.M. Douglas DC7CF fleet of aircraft. However, two of these aircraft were resold almost immediately, and the remaining two aircraft did enter service with Aer Turas. These aircraft were used on several long distance freight charter flights, and also on several charters carrying horses between the United Kingdom, Eire, Italy and France. The DC7s were also mainly operated from Shannon, although the DC4s and Bristol Freighters continued flying their services from Dublin.

For the 1972 season, Aer Turas leased an Argosy from the Canadian airline Transair - although this was Aer Turas' first turboprop aircraft, its operation did not prove to be entirely successful, and after flying several bloodstock charters between France and Eire, the aircraft was returned to Canada in September. This left the airline with one DC4 and a Bristol Freighter, and these aircraft were now almost solely used for the transportation of horses from Dublin to Luton, Cambridge, Southampton, Deauville, Le Bourget and Milan. However, the DC4 spent many months in India and Pakistan involved in relief work on behalf of the United Nations. A DC7CF entered service with Aer Turas in November 1972, and again this aircraft was mainly used for carrying horses, but was lost after overshooting the runway at Luton while landing at night on March 3 1974. This left Aer Turas with just one DC4 to operate its services, and consequently the company sought a replacement for the DC7.

A replacement was quickly found, and in April 1974 it was announced that Aer Turas had acquired a Britannia on a forty two month lease-purchase agreement from Monarch Airlines. Monarch supplied Aer Turas with maintenance for the aircraft and training for their crews, and on May 29 1974 EI-BAA entered service with Aer Turas when it operated a Dublin - Luton - Milan horse flight. During June further horse and cattle flights were made to Milan, but in July two charters were made to Africa, with one flight being made to Mauritania, and one to Lagos. Also during July, Aer Turas' Britannia flew the first two of a series of cattle flights to Greece from Dublin. During August the Britannia operated its first trans-Atlantic charter when it carried a load of horses from Stansted to New York, and at the end of the month Aer Turas flew a London - Brussels - Punta Arenas (Southern Chile) freight charter. During September a second charter to New York was flown, and a cargo charter was also made between Liverpool and Khartoum in the Sudan. In mid November 1974 EI-BAA positioned from Dublin to Abu Dhabi to start a two month charter under contract to Pauling's Middle East carrying building materials to the site of a new airfield at Midway. During this two month period, the Britannia carried nearly two thousand tons of cargo on a total of one hundred and thirty eight round trips.

The aircraft returned to Dublin in February 1975, and at the end of the month the Britannia once again left Dublin, this time to commence a series of tramping operations similar to those carried out by London Aero and Motor Services some thirty years before. During these tramping operations the aircraft made four flights from Europe to East Africa, and airports visited during these operations included Lourenco Marques, Frankfurt, Entebbe, Cairo, Rome and Kuwait. Then, early in April the Britannia was used to carry one hundred and ninety tons of cigarettes from Southend to Rotterdam. For the remainder of April and May EI-BAA flew a series of seven cattle flights from Cork to Milan, and cargo flights were also made to Khartoum, Addis Ababa, Benghazi, Sebha, Cairo and Kuwait. While the Britannia was busily engaged on these flights, the DC4 was also kept remarkably active flying numerous horse charters from Dublin to Cambridge, Deauville, Le Bourget and Luton, with the occasional flight being made to Milan during the Britannia's absence while in Abu Dhabi. A second Britannia was acquired early in September 1975 from the Royal Air Force, and after conversion by Airline Engineering at Luton, this aircraft entered service with Aer Turas late in October 1975, and made several flights between Dublin and Venice during its first month of service.

Today, Aer Turas' main source of revenue is still found in the carriage of bloodstock around Europe, and the airline is now recognised as one of the most expert in this field. Regular cattle and horse flights are now made from both Dublin and Cork to Milan, Deauville and Le Bourget, and the Britannias still make many trips to the Middle East and Africa with freight, and also to Greece with bloodstock.

Fleet:

DH89A Dragon Rapide	EI-AML	c/n 6709	dd. 6.62,	sold 5.64.	
Douglas C47 Dakota	EI-ANK	c/n 9813	dd. 4.64,	sold 1.65.	
Bristol 170 Freighter Mk.31	EI-APC	c/n 13072	dd. 3.66,	sold 11.72,	'City of Cork'.
	EI-APM	c/n 13076	dd.11.66,	written off Dublin 12.6.67.	
Douglas C54B Skymaster	EI-AOR	c/n 10441	dd. 6.65,	sold 11.69.	
Douglas DC4	EI-APK	c/n 42911	dd.12.66,	sold 7.69,	'Monarch of Munster'.
Douglas C54E Skymaster	EI-ARS	c/n 27289	dd.12.69,		'City of Galway'.
HS.650 Argosy 222	EI-AVJ	c/n 6801	lsd.11.71,	ret. 9.72,	'The Consortium', leased from Transair.
Douglas DC7BF	EI-AWO	c/n 45238	dd. 5.73,	wfu. 12.73.	
Douglas DC7CF	EI-ATT	c/n 45184	dd. 2.69,	sold 8.71,	'City of Dublin'.
	EI-ATU	c/n 45185	dd. 2.69,	sold 4.70,	'City of Limerick'.
	EI-ATV	c/n 45188	dd. 3.69,	sold 5.69,	'City of Cork'.
	EI-ATW	c/n 45186	dd. 4.69,	sold 4.69,	'City of Galway'.
	EI-AWG	c/n 45471	dd.11.72,	written off Luton 3.3.74,	'City of Limerick'.
Bristol 175 Britannia 307F	EI-BAA	c/n 12921	lsd. 5.74		'City of Dublin', leased from Monarch.
Bristol 175 Britannia 253F	EI-BBH	c/n 13436	dd.10.75		'City of Cork'.

AIKMAN

Aikman Airways Ltd. Bases: Croydon and Thame Airports.

Wing Commander Barry Thomson Aikman was Chief Navigation Officer with Transport Command for the last two years of the war, and after the war he became a director and general manager of the Lancashire Aircraft Corporation. However, in 1947 he resigned from this position, and on November 13 1947, with Cyril Walker, he registered Aikman Airways as an airline company. In February 1948, this company purchased the two Air Taxis Percival Proctors, and based at Croydon, these aircraft were joined by a Dragon Rapide in April. During the summer this fleet was used on pleasure flying and charter work, and this activity continued until the end of the year. However, Barry Aikman had left to form Aquila Airways in May 1948, and Aikman Airways had ceased charter flying by the end of 1950. By this time the airline was based at Thame airport where it had a hangar, and many former Royal Air Force aircraft were overhauled here, including several Proctors. However, several aircraft were never converted and these aircraft languished at Blackbushe and Thame for many years.

Fleet:

Percival Proctor 1	G-AGYA	c/n	K.247	dd. 2.48,	sold 11.48.
	G-AGYB	c/n	K.265	dd. 2.48,	sold 2.50.
Percival Proctor 2A	G-AKWO	c/n	H.42	ca. 4.49,	sold .50.
Percival Proctor 3	G-AKWM	c/n	H.429	dd. .49,	never converted for civil use.
	G-AKWN	c/n	H.426	dd. .49,	never converted for civil use.
	G-AKWP	c/n	H.37	ca. 5.50,	sold .50.
	G-AKWR	c/n	H.489	ca. 5.52,	sold .52.
DH89A Dragon Rapide	G-AKND	c/n	6515	ca. 4.48,	sold 4.54.

AIR ANGLIA

Air Anglia Ltd.
Anglian Air Charter Ltd.
Rig-Air Ltd.
Norfolk Airways Ltd.

Base: Norwich Airport, with outstations at Aberdeen, Humberside and Sumburgh Airports.

On July 22 1970 it was announced that three East Anglian airline and air taxi companies - Rig-Air, Norfolk Airways and Anglian Air Charter - were to merge into one unit. The new airline formed as a result was named Air Anglia, and operations began in August 1970 with aircraft varying in size from single-engined Austers to a thirty six seat Dakota. This Dakota had been used by Rig-Air on oil rig support work,and this work was continued by Air Anglia, with regular flights being operated to Aberdeen, Amsterdam, Esbjerg, Stavanger and Sumburgh. A fleet of single-engined Cessnas and Austers operated pleasure flights from Air Anglia's own airfield at North Denes, and during 1972 these aircraft were transferred to an associate company named North Denes Aerodrome Ltd. In addition, Air Anglia also operated a fleet of twin-engined Piper Aztec aircraft mainly on behalf of several locally based companies, including the Rowntree-Mackintosh and Norwich Union Assurance Groups.

When Air Anglia had been formed it was an all charter company, but on December 7 1970 the Islander G-AXVP opened Air Anglia's first scheduled service when it started a route from Norwich to Edinburgh and Aberdeen. Further expansion of Air Anglia's scheduled services came on June 29 1971 when the airline opened a service from Norwich to Liverpool and Manchester using Aztec and Islander aircraft. Two more Dakotas were acquired in 1971, mainly for use on oil related charter work from Aberdeen and Norwich - one of these aircraft was flown on an oil rig contract on behalf of Shell and Esso, and this contract mainly involved the transport of oil rig crews between Aberdeen and Sumburgh. During 1971, Air Anglia continued to operate scheduled services from Norwich to Aberdeen, Edinburgh, Liverpool and Manchester, and on December 6 1971 one of the Dakotas opened Air Anglia's third scheduled service - a daily service from Norwich to Amsterdam.

1972 was a year of rapid expansion for Air Anglia, and new scheduled services and aircraft were introduced during the year. By February 1972, four Dakotas were in service and, in addition to their scheduled service commitments, these aircraft were used during the winter of 1971/72 for transporting the Norwich City football team, and its supporters, around the country during Norwich's successful bid for promotion to League Division One. On these charter flights, Air Anglia would often use three Dakotas, and the sight of two or three of Air Anglia's aircraft lined up on many aprons around the country, became a regular weekend spectacle. In May 1972, Air Anglia inaugurated a thrice weekly Norwich - Cambridge - Jersey service which was flown in association with Intra Airways - the first flight over this route taking place on Wednesday May 10 1972. Also during May, the airline introduced its first turboprop equipment in the form of a Fokker Friendship, and this aircraft was soon placed into service on the route from Norwich to Amsterdam. Also during 1972, Air Anglia operated services at the weekends from Norwich to Rotterdam, while the services to Aberdeen were now flown via Teesside or Newcastle. Feeder services were flown daily from Cambridge to Norwich for passengers wishing to catch a connecting flight from Norwich, and a variety of aircraft were used for these link services, with the Aztecs and Twin Comanches appearing on most days. Air Anglia also signed agreements with other airlines, including Dan-Air, whereby passengers could book flights from Norwich to Kristansand and Stavanger, and on these services the passengers changed from an Air Anglia aircraft to a Dan-Air aircraft at Newcastle. The Friendship was proving very popular with Air Anglia's passengers, and on September 6 1972 this aircraft was introduced onto the Norwich - Newcastle - Aberdeen route, after spending most of the summer on the Norwich to Amsterdam route. At the end of the summer season, the services to Jersey were suspended for the winter, but the schedules to Liverpool and Manchester were closed down completely due to continuing losses.

Further expansion came in 1973, and with the introduction of a second Friendship in April, two of the Dakotas were retired from service. The surviving two aircraft were kept in service flying cargo services from Aberdeen and Norwich to Amsterdam, and also on transporting oil rig personnel between Aberdeen and Sumburgh. The Dakotas were also used on oil related charter flights to Bergen, Esbjerg, Oslo and Stavanger, and during the 1973 summer season, these aircraft were still used on scheduled services to Jersey which were no longer flown in association with Intra Airways. Intra Airways took over the operation of Cambridge to Jersey services, while Air Anglia opened direct services to Jersey from Norwich on April 18 1973, and during the summer two Friendship and two Dakota services were flown weekly in each direction. On April 21 1973, a scheduled passenger service linking Hull with Jersey was started - Leaconfield was used as the Hull terminal, but due to the lack of customs facilities here, the services were flown via Castle Donington or Norwich to clear customs. The Friendships were used on a new scheduled service from Aberdeen to Stavanger, and during the summer Air Anglia flew services from Norwich and Hull to Jersey; from Norwich to Aberdeen, Amsterdam and Teesside. Passengers wishing to travel to Newcastle were now offloaded at Teesside, and then driven to Newcastle Coach Station by coach, but this system only lasted a short time, and by 1974 the Friendships were once again stopping off at Newcastle on their Norwich to Aberdeen route.

During 1974, the route from Norwich to Aberdeen became more popular, and by the end of the year eleven flights per week were being flown over this route, with Newcastle once again becoming one of the airports visited on this service. During the summer, new services were opened from Aberdeen - in addition to direct services to Amsterdam and Stavanger, a new Aberdeen - Edinburgh - Leeds - Norwich route was also opened. However, on this latter service Air Anglia did not initially have traffic rights on the Aberdeen to Edinburgh sector, but these rights have now been obtained and the service is currently flown twice daily in each direction. One of the Dakotas, by 1974, was based at Aberdeen to operate oil rig charter flights on behalf of Shell and Esso, and Air Anglia also now based one of its many Aztecs at Aberdeen to operate air taxi flights on behalf of many oil companies. On November 11 1974, Air Anglia's Aztec G-AYTP inaugurated a twice daily service linking Norwich with Birmingham, and Aztecs were used on this service for the remainder of 1974 and much of 1975. During September and October 1974, weekly flights on Wednesdays were flown between Norwich and Blackpool carrying oil drilling support equipment. In addition, a twice weekly Dakota flight served Esbjerg from Norwich under contract to an oil company, and this service continued to be flown until the sale of the last Dakota in November 1975.

1975 was another year of rapid and historical development for Air Anglia - in February 1975 an outstation was set up at Humberside Airport, near Hull, in preparation for the introduction of scheduled services from this airport. The airfield at Humberside, previously known as Kirmington, was already a regular port of call for Air Anglia's Navajos and Aztecs on air taxi services, and one Aztec has since been based at Humberside for use on air taxi flights on behalf of locally based companies. Another Yorkshire airfield regularly visited by Air Anglia's Aztecs was the airfield at Elvington, and services were flown from here on behalf of the Rowntree-Mackintosh Group who have factories in nearby York, On April 1 1975, a twice every weekday Edinburgh to Amsterdam service was opened, and on April 7 Air Anglia opened its first scheduled services from Humberside when one of the Friendships inaugurated a weekday service to Amsterdam. A Humberside to Jersey scheduled service was opened by Air Anglia's leased Herald (G-ASVO) on April 19 1975, and this route was flown at weekends by Air Anglia throughout the summer, replacing the Leaconfield to Jersey service. At this time a service from Aberdeen to Jersey was also opened - this becoming Air Anglia's longest distance direct scheduled service by Friendship.

During the summer, two Dakotas remained in service and, although they were mainly used for cargo flights, these aircraft remained extremely active. Apart from the twice weekly Norwich to Esbjerg service already mentioned, the Dakotas flew scheduled cargo services from Aberdeen and Norwich to Amsterdam. The Dakota used on this service would fly from Amsterdam to Norwich early in the morning, then return with another load of cargo to Amsterdam early in the afternoon, before making the long over sea trek to Aberdeen in the early hours of the evening, while the return flight to Amsterdam was made late that evening. These services were flown daily by Air Anglia, and fully utilised one aircraft for the year. Air Anglia's 'bus stop' services from Norwich to Aberdeen were so popular during the height of the summer that an extra Friendship was acquired in June, and at peak times the airline operated direct Norwich to Aberdeen services. The intensive nature of Air Anglia's operations has meant that few charter flights are undertaken by the Friendships during the summer, but on Sunday October 5 1975 one of these aircraft operated a charter flight from Norwich to Billund.

The summer services to Jersey were suspended early in October, but the end of the summer did not mean the end of any further expansion for Air Anglia. On October 27 1975, a twice weekly Norwich to Stavanger service was opened - prior to this, passengers wishing to travel from Norwich to Stavanger had to land at Leeds, Edinburgh and Aberdeen en-route. Thus, with the introduction of this new direct service, not only was the service quicker, but it also released more seats on the routes from Norwich to Aberdeen for other passengers. In November 1975 the Norwich to Birmingham service became a Navajo Chieftan service, and in order to operate this route, a second hand aircraft was bought in North America.

On November 24 1975, the last Dakota was retired from service, and on this date Air Anglia introduced its first four engined aircraft in the form of an Argosy. This aircraft has more than doubled Air Anglia's available cargo capacity on its scheduled cargo services from Aberdeen and Norwich to Amsterdam. Initially, the company operated one of ABC's aircraft on lease, but on December 6 Air Anglia took delivery of G-APRL in its own colours, and this aircraft has been obtained on a twelve month lease from Field Aircraft Services. After five years of Dakota operations, Air Anglia had now retired all of these aircraft and, apart from the Norwich to Birmingham service, all scheduled services were now flown by turboprop aircraft. Further expansion came on November 19 1975 when, on this date, Friendship G-BCDN positioned from Norwich to Leeds to open a Leeds to Amsterdam service. Air Anglia took over this route from Northeast, who had been losing money on the service for some years, and Air Anglia now operate this service thrice weekly either with their own Friendships or leased Alidair Viscounts. The licence for this route, though, has not been awarded yet to Air Anglia, and Dan-Air have also applied to operate over this route.

For the future, Air Anglia holds licences for an Aberdeen - Isle of Man - Swansea service, and also for an extension of the Norwich to Birmingham service to Swansea. Air Anglia is one of Europe's fastest growing air-lines, carrying well over one hundred thousand passengers every year, and a quick perusal of the airline's passenger figures soon proves to any sceptic that Air Anglia's bold venture in 1972 to expand its scheduled services with the introduction of Friendship aircraft, has paid off handsomely. In the year 1971/72, a total of twenty three thousand passengers were carried by the airline; in 1972/73 this total had increased to forty eight thousand, and for the year 1973/74 to a total of over seventy five thousand passengers.

Fleet:

Auster J5G Autocar	G-ASFK	c/n 3276	ISOF	transferred to North Denes Aerodrome Ltd. 7.72
Cessna 170B	G-AWOU	c/n 25829	ISOF	transferred to North Denes Aerodrome Ltd. 7.72
Cessna F.172H	G-AVEC	c/n 0405	ISOF	sold 8.71.
	G-AVIE	c/n 0326	ISOF	
	G-AVVZ	c/n 0475	ISOF	sold .72.
Cessna U.206A	G-ATLT	c/n 0523	ISOF	transferred to North Denes Aerodrome Ltd. 8.72
Douglas C47A Dakota	G-AGJV	c/n 12195	dd. 2.72,	sold 10.75.
Douglas C47B Dakota	G-AMPZ	c/n 32872	ISOF	sold 2.73.
	G-ANTD	c/n 25414	dd. 4.71,	wfu. 4.73.
Douglas C53D Dakota	G-AOBN	c/n 11711	dd.11.71,	sold 11.75.
BN.2A Islander	G-AXVP	c/n 127	ISOF	sold 3.73.
Piper PA27 Aztec 250C	G-BAAJ	c/n 27-2569	dd. 7.72,	sold 10.74.
Piper PA27 Aztec 250D	G-AXDC	c/n 27-4169	ISOF	
	G-AYBO	c/n 27-4510	ISOF	
	G-AYSF	c/n 27-2569	dd. 2.71	
Piper PA27 Aztec 250E	G-AYTP	c/n 27-4585	dd. 3.71	
	G-BCBG	c/n 27-5224	dd. 4.74	
Piper PA31 Navajo 300	G-BBML	c/n 31-1210	dd. 3.74	, operated for Norwich Union.
	G-BBZI	c/n 31-1211	dd. 3.74	, operated for Reckitt and Colman.
Piper PA31 Navajo 350	G-BBMD	c/n 31-5119	dd.11.75.	
Piper PA34 Seneca 200	G-AZON	c/n 34-50081	lsd. 73, ret. .73.	
Piper PA30 Twin Comanche	G-ATEN	c/n 30-544	dd. .71,	sold 5.72.
Piper PA39 Twin Comanche C/R	G-AYZY	c/n 39-89	dd. 6.71,	sold 3.73.
Fokker F.27 Friendship 200	G-BAKL	c/n 10293	dd. 5.72,	operated as PH-FIL during 1972, 'Phil'.
	G-BAUR	c/n 10225	dd. 4.73,	, 'Bess'.
	G-BCDN	c/n 10201	dd. 5.74	
	G-BCDO	c/n 10234	dd. 6.74	
	G-BDDH	c/n 10289	dd. 6.75	
Fokker F.27 Friendship 400	PH-FPR	c/n 10442	lsd.12.72, ret. 1.73,	leased from Fokker.
Handley Page Herald 214	G-ASVO	c/n 185	lsd. 3.75	, leased from British Midland.
AW.650 Argosy 102	G-APRL	c/n 6652	lsd.12.75	, leased from Field Aircraft Svs.
	G-AZHN	c/n 6657	lsd.11.75, ret. 12.75,	leased from Air Bridge Carriers

AIRBORNE TAXI SERVICES

Airborne Taxi Services Ltd. Base: Elstree Aerodrome.

Airborne Taxi Services was registered on December 10 1946 as an airline company by Wing Commander O.V.Holmes, Major R.R.W.Jackson and Captain J.A.F.Miers of Bracknell in Berkshire. The company started light charter operations from Elstree Aerodrome with a fleet of single engined aircraft, and over the next six years Airborne Taxi Services operated a varied fleet of Cub, Hornet Moth and Auster type aircraft, on short range ad-hoc charter flights. The company offered its aircraft for charter from Lympne Aerodrome and several cross-Channel flights were made from Lympne to airfields in northern France, while pleasure flying was carried out at Lympne during the summer months. Other work undertaken by Airborne Taxi Services included a series of army co-operation flights undertaken during 1949 and 1950. Wing Commander Holmes became the company's chief pilot, and an office was later opened in Holland Park Avenue, London. The company also used an Airspeed Consul during 1950 and 1951, but details of this aircraft seem rather scarce, and it is not known which particular aircraft was used by this company. When Airborne Taxi Services stopped flying late in 1952, the company was operating a Hornet Moth and a Taylorcraft mainly on light charter work and joy riding activities, particularly at airshows during the summer months.

Fleet:

Piper L.4 Cub	G-AKAA	c/n 10780	dd. 8.47,	sold .50.
Auster 5	G-AKTA	c/n 1319	rd. 5.51,	sold 6.51.
Taylorcraft Plus D	G-AHXF	c/n 184	dd. .50,	sold 11.52.
DH87B Hornet Moth	G-AELO	c/n 8105	dd. .46,	ISOC.
Airspeed AS65 Consul	G-A	c/n	dd. .50,	sold . .

AIR BRIDGE CARRIERS

Air Bridge Carriers Ltd. Base: East Midlands Airport.
Carbia Transport Ltd.
Associated British Carriers Ltd.

When Sagittair ceased operations in August 1972, three Argosies, which Field Aircraft Services had been responsible for maintaining, were left without work. Therefore, a new company named Air Bridge Carriers was formed as an associate company to Field Aircraft Services. Initially licence applications, to take over Sagittair's services, were made in the name of Carbia Transport, but in November 1972 the first Argosy entered service in the colours of Air Bridge Carriers. Warehousing facilities were obtained at Castle Donington, and during the spring and summer the Argosy, later joined by two others, plies between Castle Donington and Jersey and Guernsey to collect flowers, fresh fruit and vegetables. The flowers after arrival at Castle Donington, are distributed to markets in the south and to Leeds, Derby, Newcastle, Preston and many other markets in the Midlands and the north. The aircraft normally carry a load of 2,000 boxes of flowers, with seventy two delicate blooms packed in each box. The licence for these flights runs for twenty years - and outward from Castle Donington to Guernsey and Jersey the aircraft often carry other freight such as fish, or various goods for shops in the Channel Islands. At the height of the season, the airline flies up to eight services a week between Castle Donington and the Channel Islands. Some services are flown to Bournemouth, and although one Argosy was sold in April 1974, a Viscount was acquired during the same month, and this aircraft entered service late in September 1974. During the winter of 1974 the airline operated numerous ad-hoc freight charters from Castle Donington, and some oil rig work was carried out from Aberdeen. Recent charters have taken the aircraft as far as Zagreb, and regular charters are made to many European cities carrying freight including the carriage of racehorses.

In February 1975 the Viscount was converted to a convertible passenger/cargo aircraft, and during the year it operated Dan-Air's cross-Channel coach-air services between Lydd and Beauvais. During April 1975, one of the Argosies opened Alidair's twice nightly freight service between Liverpool and Belfast, but this service was taken over by Alidair on May 1. The first of these services was flown by A.B.C.'s Argosy G-AZHN on April 2 1975. During 1975 several flights were operated on behalf of Rolls Royce, with many flights from France and Germany to Filton, carrying aero engines and spares. Other services have been flown from Italy and West Germany to Warton on behalf of the British Aircraft Corporation and Rolls Royce, and early in November 1975 one of the Argosies (G-APRN) was painted up as a support aircraft for a tour of the Middle East as support aircraft for the B.A.C. Jaguar. The Argosies have also carried out several charters between Castle Donington and Rennes, in France carrying cows and calves, but a shortage of aircraft in October and November 1975 led to some flights being subchartered to British Air Ferries.

Fleet:

HS.650 Argosy 102	G-APRL	c/n 6652	dd.11.75	
	G-APRN	c/n 6654	dd. 6.73	
	G-APWW	c/n 6656	dd.11.72,	sold 4.74.
	G-AZHN	c/n 6657	dd. 8.73	
Vickers 808 Viscount	G-BBDK	c/n 291	dd. 4.74	

AIR CARGO DISTRIBUTORS

Air Cargo Distributors Ltd. Base: Hatston Aerodrome Kirkwall.

Air Cargo Distributors was a company formed in February 1948 to operate cargo flights between the Orkneys, Shetlands and Woodley. The company initially operated a Miles Aerovan, on lease from Kennings, based at Hatston Aerodrome. These services were principally intended as a rapid outlet for products and commodities of the islands, and from Woodley the aircraft returned to the Orkneys and Shetlands with cargoes of food, vegetables and items of general merchandise. Some of the local products of the Orkneys and Shetlands carried by the Aerovan included lobsters for markets in the south. The founders of this airline were Sir Basil Neven Spence, M.P.(who acted as Chairman), Colonel J.E.Henderson and Mr.A.Mackay. London directors were Mr.H.C.Martineau (who was also a director of International Exporters and Importers Ltd.), Mr.M.Adams and Squadron Leader J.M.Clark. In February 1949 the leased Aerovan was replaced by one of Air Cargo's own aircraft. During the next eighteen months, the airline operated various ad-hoc charters from both Kirkwall and Woodley, but in October 1950 the aircraft, which was owned by the Avia Britannica Corporation, was sold and all operations came to an end.

Fleet:

Miles M.57 Aerovan 4	G-AJZP	c/n 6421	lsd. 2.48,	ret. 3.49,	leased from Kenning Aviation.
Miles M.57 Aerovan 6	G-AKHF	c/n 6399	dd. 2.49,	ISOC ,	'Northern Exporter'.

AIR CHARTER

Air Charter Ltd. Bases: Croydon and Bovingdon, and later Stansted, and Southend.

Early in 1947 Air Charter was formed by Mr.Freddie Laker as a charter operator based at Croydon, and passenger and freight ad-hoc charter operations started in March 1947 using a Dragon and a Dragon Rapide. These two aircraft had been joined by a second Dragon Rapide and two Miles Geminis by the end of the year, and during the next three years these aircraft formed the backbone of Air Charter's fleet. During 1947 Laker also set up a maintenance organisation known as Aviation Traders. Based at Southend, this company during the ne t three years purchased large numbers of war surplus aircraft, most of which were either converted for civilian use or were broken up for spares. When the Berlin Airlift started, Laker took no direct part in the airlift, but instead he purchased the B.O.A.C. fleet of Haltons, converted them, and many were then used on the Berlin Airlift by Bond Air Services. When the aircraft returned to Southend at the end of the airlift in August 1949, the aircraft were withdrawn from service and scrapped. The depression in the air charter market after the airlift caused Laker to turn to scrapping aircraft and melting them down, and no charter work was undertaken, resulting in the sale of the Rapides and Geminis during 1950. Air Charter's return to charter work came early in 1951 through one of its associate companies - Mr.Laker's Aviation Traders took over Surrey Flying Services early in 1951, and in May 1951 this company restarted passenger and freight charter operations with a York. By November 1951 this company was operating two Yorks and one Tudor, and during this month Surrey Flying Services took over Fairflight - a Blackbushe based Tudor operator. Fairflight held a contract to carry freight from Berlin to Hamburg, and it is this contract which Air Charter took over, and upon which the airline would base its growth over the next three years.

The combined fleet of aircraft at November 1951 stood at two Tudors, two Yorks and one Lincoln, and these aircraft were all used on the lucrative freight run between Berlin and Hamburg. In July 1952 both Surrey Flying Services and Fairflight dropped their titles, and operations continued under the Air Charter flag. Surrey Flying Services had already moved their base to Stansted which had become Air Charter's base of operations, and although Fairflight's Tudor acquired Air Charter titles, the Lincoln was withdrawn from service and scrapped. Although Surrey Flying Service's name had disappeared from its aircraft, the name continued to be used for some time afterwards, and the company successfully tendered for several Air Ministry freight and trooping contracts using its Avro Yorks. On these flights the Yorks plied regularly into and out off Lyneham, Abingdon and several other Royal Air Force Transport Command bases. However, the second airlift into Berlin provided the airline with the majority of its work - this airlift was started due to difficulties with the Russians interfering with the transport of goods out of West Berlin, and this type of air service once again seemed to provide the answer. Initially two Yorks were based in Hamburg to operate these flights, and by Septmber 1952 Air Charter was making up to four round trips daily using these Yorks, each time carrying eight tons of freight from Berlin to Hanover and Hamburg. Outward bound from Berlin the aircraft carried finished products, banned by the Russians, and inward the aircraft carried raw materials such as copper ingots and graphite which might not have otherwise successfully completed their journey from West Germany to Berlin along the autobahns in Russian held territory. By November 1952 the number of flights operated had been increased to seventy flights per week - between Hanover and Hamburg, some twenty three passenger trips and forty seven freight runs were made during the first week of November in addition to the freight flights to Berlin. In February 1953 the Yorks were joined by a Bristol Freighter, followed by a second aircraft exactly a year later.

In February 1954 the first of many Avro Supertraders entered service with Air Charter. When Laker acquired Fairflight's Tudor in 1951, he looked into the possibility of obtaining more of these aircraft for use on Air Charter's trooping and freight services from Stansted and Hamburg. After purchasing virtually every available airframe, Aviation Traders undertook a massive conversion programme culminating in the aircraft's entry into service on February 14 1954, when G-AGRG flew Air Charter's first Supertrader service from Stansted to Hamburg and return. During the next two years the majority of the Yorks were sold as more Supertraders entered service. One of the Tudor 4s (G-AHNM) was flown to Berlin in March 1954 to assist in flying the cargo services from Hamburg. This aircraft carried over nine tons of cargo on each flight, and operated at an all up weight some 5,000 lb below the permitted maximum of 82,000 lb. The Tudors continued to operate the Hamburg to Berlin freight run, and also operated numerous trooping flights to the Middle and Far East with servicemen and their families, and later a freight service was flown to the Woomera rocket range in Australia on behalf of the British Government. The Tudors also flew a colonial coach class passenger service from Stansted to Idris and Lagos with the sircraft in a forty two seat configuration.

Air Charter's first Douglas DC4 arrived at Southend in February 1955, and after conversion was based in Hamburg for use on the freight runs, and also on the passenger services radiating from Berlin. Air Charter opened their Southend to Cyprus trooping run with DC4s during August 1956, and later more trooping runs were flown from Southend with the DC4s. Prior to this, on April 14 1955 Air Charter started vehicle ferry services between Southend and Calais using Bristol Superfreighters. These services proved to be extremely popular, and in June 1957 a joint service with SABENA from Southend to Ostend was started. In 1956 some 8,500 cars were carried on these services, and by 1958 this figure had increased to 13,000, in addition to the 63,727 passengers carried on these scheduled services during that year. In 1957 the freight services radiating from West Berlin came to an end, but the aircraft returned to the United Kingdom and soon found other work on Ministry of Defence contract work. During 1958 Air Charter's operations continued unchanged - the Tudors continued plying to Australia and the Far East on their Ministry contracts, and on the Government service to the Woomera rocket range - the aircraft operated these services from Stansted, and their first stop outbound was Brindisi. The aircraft were often away from their base for up to two weeks, and the Tudors were rarely used for any other work. The Bristol Freighters, though, were occasionally used for ad-hoc freight charters, including the carriage of racehorses, and these charters included flights from Stansted and Gatwick to Caen and Le Bourget.

A replacement for the Tudors was sought for during 1958, and on September 12 1958 the first of two Britannias entered service on Air Charter's trooping runs from Stansted. On October 1 1958 Britannia G-ANCE made the first of several trooping runs between Stansted and Christmas Island with up to one hundred and twenty four passengers per flight. During the next two months this aircraft became a familiar sight at many airports abroad including Goose Bay, Montreal, Gander and Keflavik in the hands of Captains Hellary, Smith and Jennings. On February 5 1959 the Britannia left Stansted carrying one hundred service personnel to Adelaide and Sydney. The aircraft arrived in Australia on February 8, and then flew to Christmas Island to collect one hundred and twelve more passengers, who were then flown back to Stansted via San Francisco and Montreal, thus completing a round the world flight in a flight time of seventy two and a half hours - the aircraft touched down at Stansted on February 12. The aircraft left Stansted again on February 18 bound for Istanbul and the Far East, before returning to Stansted via Brindisi on February 23.

While the airline was involved with the introduction of this new type of aircraft, the Airwork group of companies had started negotiations with Air Charter late in 1958 with a view to the take-over of Air Charter by Airwork. Thus it was that in January 1959 Air Charter became a subsidiary of Airwork, and although Air Charter continued to operate under its own name, there was a rationalisation of crews and ground staff. In February 1959 the vehicle ferry division at Southend became a separate company when a new airline named Channel Air Bridge was formed to take over the fleet of Bristol Freighters and the vehicle ferry services. 1959 also saw many other changes - the introduction of the Britannias heralded the retirement of the now ageing fleet of

EI-AFA was one of Aer Lingus' many Dakotas - this particular aircraft served with the airline for fifteen years.

Ian MacFarlane

Aer Lingus' first turboprop aircraft were Viscount 700's, which were later replaced by a fleet of larger series 800 Viscounts - EI-AJI is illustrated here.

Vickers Ltd.

Air Charter's Tudor G-AHNI completed its last flight at 0843 on June 10, 1959, when Captain Treharne landed this aircraft at Stansted after a flight from Benghazi.

A J Jackson

Aer Turas used this Bristol Freighter EI-APC for many flights carrying British, French and Irish racehorses around Europe.
Transportraits/John Goring

Aer Turas' Dakota EI-ANK, pictured here at Newcastle, flew to Cardiff on June 18, 1964 to have the aircraft weighed by Cambrian Airways.
Ian MacFarlane

EI-BBH, Aer Turas' second Britannia, entered service on October 28, 1975 after conversion by Airline Engineering at Luton.
Tony Merton Jones

G-AXDC Air Anglia Aztec D — Tony Merton Jones

G-AZHN ABC Argosy 102 — Austin J Brown

G-ANMF Air Charter Freighter 31 — Ian MacFarlane

G-ANVR Air Charter Freighter 32 — Transportraits/John Goring

G-ANYB Air Charter Douglas DC-4 — Ian MacFarlane

G-AYEM Air Charter Scotland Aztec D — Ian MacFarlane

Tudors, and two fatal accidents early in 1959 with this type of aircraft speeded its retirement. The Tudors were retired from the runs to the Woomera rocket range at the end of April 1959, and during May the surviving aircraft (G-AHNI and G-AHNL) made several trips to Benghazi, Malta and Benina. But the end of the Tudor's commercial career was fast approaching, and on June 10 1959 Captain Treharne landed G-AHNI at Stansted at 08.43 after a flight from Benghazi. The other Tudor was to make one last flight, and on June 15 G-AHNL left Stansted for Lisbon in the hands of Captain Alford, and when the aircraft touched down at Stansted on June 16 1959 at 14.41 after its return flight from Lisbon, this marked the last flight in the world of any variant of the Avro Tudor.

The retirement of the Tudor meant that DC4s appeared at Stansted on many of the Ministry of Defence contract services to the Middle East. Before this the DC4s had rarely visited Stansted apart from the occasional crew training detail or ad-hoc charter, but on April 16 1959 G-AOXK left Stansted for Brindisi with Captain Couch in command on its way to Australia, returning two weeks later on April 30. After this the DC4s became regular visitors to Stansted on these services, and during the year these aircraft also opened up other new services. On March 10 1959 one of the airline's DC4s (G-ANYB), and a Bristol Superfreighter (G-ANVS) of the associated Channel Air Bridge, inaugurated a new service between Southend and Calais whereby passengers would be allowed to travel to France for day trips with no passports. On the first flights, mainly the Press were carried, but during the summer this service proved to be extremely popular, and if the passengers preferred one of the journeys could be made by sea. Another new service inaugurated by the DC4s during 1959 was a service from Australia to Stansted transporting Jindiviks to this country from mid 1959 onwards.

Although Air Charter's second Britannia was delivered to the airline in March 1959, it did not enter service until later in the year. The aircraft was flown to Southend for painting and the fitting of certain equipment, and it did not arrive at Stansted until May 25. On the following day the aircraft carried out an air test at Stansted, and on May 27 Captain Lane lifted G-ANCD from the runway at Stansted early in the morning for a flight to Lisbon. The aircraft returned to Stansted during the afternoon, but left again for Lisbon early in the evening, and the return journey to Stansted was made on May 28. Thus, Air Charter now had two Britannias in service , and during the next few months these aircraft were to make several interesting flights - on June 26 1959 G-ANCE flew direct from Karachi to Stansted, while on July 4 the same aircraft left Stansted for Leopoldville in the Belgian Congo to bring Europeans back to Brussels, before returning to Stansted on July 8. Air Charter's fleet modernisation was now complete, and when the company became part of Britsh United Airways late in June 1960, the airline was able to boast a fleet of two Britannia, four DC4s, and Channel Air Bridge's fleet of Bristol Freighters. Air Charter's managing director, Mr.Freddie Laker, was to become managing director of British United Airways.

Fleet:

Airspeed AS.65 Consul	G-AJGH	c/n 5124	dd. 4.48,	sold 6.50.	
DH84 Dragon 2	G-ADDI	c/n 6096	dd. 3.47,	sold 2.51.	
DH89A Dragon Rapide	G-AFHY	c/n 6417	dd. 3.47,	sold 6.50.	
	G-AGFU	c/n 6463	rd. 8.47,	sold 6.50.	
Miles M.65 Gemini 1A	G-AJWL	c/n 6460	dd. 8.47,	sold 6.50.	
	G-AKGA	c/n 6474	dd. 9.47,	sold 6.50.	
	G-AKDD	c/n 6284	dd. 9.55,	sold 9.57.	
Avro 685 York 1	G-AMGL	c/n 1354	dd.11.51,	written off	Hamburg 11.3.52, 'New Era', trooping as XA192.
	G-AMGM	c/n 1355	dd.11.51,	written off	Lyneham 27.11.52, 'New Venture'.
	G-AMRI	ex. MW138	dd. 8.52,	sold 6.55,	'New Enterprise'.
	G-AMRJ	ex. MW326	dd. 5.52,	sold 6.55,	'New Era II', trooping as XG897.
	G-AMUS	ex. MW110	dd.10.52,	sold 2.56,	'New Britain', trooping as XF919.
	G-AMUT	ex. MW185	dd.10.52,	sold 3.55.	
	G-AMUU	ex. MW183	dd.10.52,	sold 2.56,	'Nouvelle Caledonie', trooping as XD668.
	G-AMUV	ex. MW226	dd.10.52,	sold 3.56,	'New Venture II', trooping as XD623 and XD669.
	G-AGNU	c/n 1222	dd. 9.52,	sold 12.56,	'New Endeavour', trooping as XD670
	G-ANXN	ex. MW258	dd.12.54,	sold 11.56,	'New Charter'.
Avro 688 Tudor 1	G-AGRI	c/n 1257	dd. 2.54,	wfu. 10.54,	trooping as XF739.
	G-AGRJ	c/n 1258	dd. 8.54,	wfu. 8.56.	
	G-AIYA	c/n 1367	dd. 9.54,	wfu. 5.55.	
Avro 688 Tudor 2	G-AGRY	c/n 1262	dd.11.51,	wfu. 7.59,	trooping as XF537.
Avro 688 Tudor 3	G-AJKC	c/n 1368	dd. 9.54,	wfu. 8.56.	
Avro 688 Tudor 4B	G-AGRG	c/n 1255	dd. 2.54,	written off	Brindisi 27.1.59, 'El Alamein'.
	G-AGRH	c/n 1256	dd. 7.56,	written off	Turkey 23.4.59, 'Zephyr'.
	G-AHNI	c/n 1342	dd. 4.55,	wfu. 6.59,	'Trade Wind'.
	G-AHNL	c/n 1345	dd. .55,	wfu. 6.59.	
	G-AHNM	c/n 1346	dd. 3.54,	wfu. 6.59,	'Cirrus'.
	G-AHNO	c/n 1348	dd. 7.55,	wfu. 8.59,	'Conqueror'.
Douglas C54B Skymaster	G-ANYB	c/n 10528	dd. 2.55,	ISOM,	'Atlanta'.
Douglas C54A Skymaster	G-AOFW	c/n 10351	dd.12.55,	ISOM,	'Jason'.
Douglas DC4	G-AOXK	c/n 42931	dd.10.56,	ISOM,	'Golden Fleece'.
Douglas C54B Skymaster	G-APNH	c/n 18333	dd. 3.60,	ISOM.	
Bristol 170 Freighter Mk.31	G-AMLP	c/n 13078	dd. 2.53,		'Vanguard', converted to Mk.32 March 1958.
	G-AMSA	c/n 13142	dd. 2.54,		'Voyager', trooping as XH385, Converted to Mk.32 July 1958.
	G-ANMF	c/n 13216	dd. 8.54,		'Victory'.
Bristol 170 Freighter Mk.32	G-ANVR	c/n 13251	dd. 3.55,		'Valiant'.
	G-ANVS	c/n 13252	dd. 4.55,		'Vigilant'.
	G-AOUU	c/n 13257	dd.12.56,		'Venture'.
	G-AOUV	c/n 13258	dd.12.56,		'Valour'.
	G-APAU	c/n 13256	dd. 6.57,		'Versatile'.
	G-APAV	c/n 13263	dd. 4.57,		'Viceroy'.

All Bristol Freighters were tranferred to Channel Air Bridge in February 1959, whose history appears elsewhere.

Bristol 175 Britannia 307	G-ANCD	c/n 12920	dd. 3.59,	ISOM.
	G-ANCE	c/n 12921	dd. 9.58,	ISOM.

AIR CHARTER EXPERTS

Air Charter Experts Ltd. Base: Ronaldsway, Isle of Man.

Air Charter Experts was formed in March 1947 by Captain G.H.Drummond, who acted as chairman, Mr.L.Vondy and Mr.W.M.Finch. Two veteran Dragon Rapides were acquired to operate charter flights to England, Scotland and Northern Ireland, and the company also ran an associated aero club. However, after little over five months of charter flying Air Charter Experts was taken over by Manx Air Charters in August 1947. Only one of the Rapides was taken over by Manx Air Charters, while the other aircraft was withdrawn from service and broken up at

Ronaldsway during 1948. The name of Air Charter Experts disappeared, but its associated flying club continued under the ownership of Manx Air Charters.

Fleet:

DH89 Dragon Rapide	G-ADAE	c/n	6272	dd. 5.47,	ISOT.
	G-AEMH	c/n	6336	dd. 3.47,	ISOT.

AIR CHARTER (SCOTLAND)

Air Charter (Scotland) Ltd. Base: Glasgow Airport.

This air taxi company was founded in October 1965 by Captain Geoff Rosenbloom - formerly chief pilot and managing director of Assistair. The company acquired an Airgo Aztec during October, and started air taxi and executive charter flights from Glasgow during the same month. The company's insignia, a Red Lion, is carried on the port side of the Aztec's tail only, and the Aztec is used on many charter flights connected with the search for oil off the east coast of Scotland, and the aircraft has visited many airfields in Scotland and Northern England including Edinburgh, Aberdeen, Sumburgh and Newcastle on charter flights. Charters are also flown to many towns in Scandinavia from airports in Scotland.

Fleet:

Piper PA27 Aztec 250D	G-AYEM	c/n 27-4411	dd.10.75.

AIR COMMERCE

Air Commerce Ltd. Base: Croydon Airport.

Air Commerce was formed early in 1937 by British and Foreign Aviation, and during 1938 scheduled passenger services were opened linking Liverpool with London, Belfast and the Isle of Man. Although operations were suspended on the outbreak of war in 1939, these scheduled services were reopened in 1945. In December 1946, the airline was taken over by Olley Air Service, and scheduled services continued until February 1 1947 when all the airline's scheduled routes were taken over by B.E.A. Air Commerce's name continued in existence, however, until 1950 as a charter operator using Olley Air Service's aircraft, and operating these services from Olley's base at Croydon.

Fleet:

DH89A Dragon Rapide	G-AERZ	c/n	6356	dd. 5.37,	written off County Down 1.4.46.
	G-AFEP	c/n	6406	dd.11.40,	sold 12.46.
	G-AHGG	c/n	6902	dd. 8.46,	sold 12.46.
DH90 Dragonfly	G-AEFN	c/n	7507	dd. 5.39,	sold 3.40.

AIR COMMUTER

Air Commuter Ltd. Base: Coventry Airport.

Air Commuter was formed at Coventry Airport late in 1969, and charter operations began in January 1970 with one Dove aircraft. The first work for the Dove was flying under contract to Richard, Johnson and Nephew Ltd., flying that company's executives around the British Isles while their own aircraft was undergoing an overhaul and repaint. The Dove later returned to Coventry once it had completed this work, but later in the year the aircraft positioned to Norwich to operate several executive charter flights on behalf of Tyler Aviation. Most of 1970 was, therefore, spent flying charters on behalf of other companies, but during the year Air Commuter opened an office at Coventry Airport and started to fly many ad-hoc passenger and freight charters within the British Isles and Europe. This work continued throughout 1971 and 1972, but early in November 1973 a Piper Seneca arrived at Coventry Airport for use by Air Commuter. The aircraft is registered in the name of Escadale Properties Ltd., but Air Commuter use this aircraft on their air taxi flights from the Midlands, and also operate the aircraft on behalf of its owners. The Dove was sold during 1974, but is still based at Coventry and regularly commutes to Cork in Southern Ireland in the hands of its owner, Mr.J.Filhol.

Fleet:

DH104 Dove 6	G-ASNG	c/n 04485	dd. 1.70, sold .74.
Piper PA34 Seneca 200	G-BBNH	c/n 34-50339	dd.11.73.

AIR COMPTON

Air Compton Ltd. Bases: Blackbushe and Compton Abbas.
George House Aviation Ltd.

In July 1972 George House Aviation started air taxi and executive charter operations from Blackbushe Airport with one Piper Aztec - the company was formed as a subsidiary of the George House (Holdings) Group. The Aztec was used on a variety of air taxi services, and several flights were made between Blackbushe and Manchester. In June 1975 this aircraft was joined by a second Aztec, and during the summer the company bought Compton Abbas airfield in Dorset, and on August 1 1975 the airfield was officially reopened. George House Aviation changed its name to Air Compton, and at the same time moved its base to Compton Abbas. At this airfield a flying school was set up named the Air Compton Flying Centre, and the company provided residential and bar facilities for club members. In addition, Air Compton set up maintenance and hangarage facilities for its fleet of aircraft at Compton, and the airfield is now open to visiting light aircraft. Today the two Aztecs are still used for air taxi flights, and operating from both Compton Abbas and Blackbushe they have operated a variety of air taxi flights during 1975. In addition to the normal run of domestic air taxi flights, one of the Aztecs (G-AZXG) operated an ambulance flight from Lydd to Zurich and return on May 26 1975, and since then the company has operated numerous flights from Rochester to Warton, Woodford and Filton on behalf of Elliot Automation. Other charters have taken Air Compton's aircraft to destinations including Antwerp, Manchester and Perranporth.

Fleet:

Piper PA27 Aztec 250D	G-AZXG	c/n 27-4328	dd. 7.72.
Piper PA27 Aztec 250E	G-BBNK	c/n 27-5119	dd. 6.75.

AIR CONDOR

Air Condor Ltd. Base: Southend Airport.
Shortcut Aviation Ltd.

Air Condor was formed at Southend Airport early in 1960 by Mr.R.A.Short and Mr.Cutler to operate passenger and freight charter flights. This followed the closing down of Shortcut Aviation, which Air Condor's two founders had formed with a Bristol Freighter in 1958. This firm had been based at Blackbushe Airport, and Air Condor's first aircraft was the Bristol Freighter previously flown by Shortcut Aviation. This aircraft (G-AGPV) first flew in Air Condor's colours on March 21 1960, having been owned by Shortcut Aviation since November 1958. The Brsitol Freighter was used initially for carrying livestock, including pigs, from England to Dublin. An associate company of Air Condor, Skycraft Services, operated charter flights from Dublin, and was also responsible for the ground handling of many airlines' aircraft at Dublin. The Bristol Freighter also undertook passenger charters including several trips to Amsterdam, Ostend and Rotterdam, and a charter from Rotterdam to Hurn and return on April 14, but eight days later, on April 22 1960, the aircraft was seriously damaged when it swung on take-off at Southend and hit an Auster. The aircraft was then temporarily repaired, and flown to Hurn for complete repairs to be carried out by Airwork Services.

The loss of this Bristol Freighter left Air Condor with no aircraft to fly its charter programme, and in order to carry on flying the airline leased another Bristol Freighter from North South Airlines, and also a Vickers Viking from Keegan Aviation. During the spring Air Condor's Directors had successfully negotiated several important inclusive tour charter contracts and during the summer months this activity was to keep the aircraft busy flying holidaymakers to destinations all over Europe. In addition to many flights from Southend to several continental destinations the Viking and Bristol Freighter (in passenger configuration) were regular visitors to Manchester, carrying tourists to Ostend several times each week throughout the summer months. At the height of the season, the two aircraft were almost daily visitors to Manchester, particularly at the weekend. An occasional Rotterdam to Manchester service was also flown, and during July the Viking made several trips to Dublin from Gatwick and Manchester - on July 20 G-AKTU flew an early morning freight run from Manchester to Dublin and the Isle of Man. During July the Viking also flew several charters from Gatwick to Amsterdam and Ostend, and also a charter from Manchester to Basle on July 27. The company also occasionally ran holiday charter flights from Renfrew Airport, Glasgow - the Bristol Freighter would position from Southend to Glasgow on the Friday evening, and on the Saturday it would fly from Glasgow, via Birmingham, to Jersey and return, before flying back to Southend on Sunday morning. An Amsterdam to Glasgow charter flight was also flown on a few occasions, but from Gatwick the airline often flew services using the Viking. During the summer months this aircraft flew charters to Amsterdam, Basle, Dublin, Le Bourget, Lyons, Ostend, Rotterdam and Tours.

Air Condor's own Bristol Freighter (G-AGPV) was repaired by Airwork Services at Hurn, and test flew on July 29 1960. The aircraft was then made ready for delivery to Air Condor on July 30, but technical problems delayed its delivery until early the following morning when the aircraft positioned to Southend. Its first day back with Air Condor was a busy one, flying a charter to Amsterdam, then from Amsterdam the aircraft flew to Glasgow, before returning to Southend in the evening. On the following day, August 1, the aircraft flew to Jersey, then to Glasgow, and again returned to Southend in the evening. The remainder of the season saw the continuation of the Manchester to Ostend services, and Air Condor's many charter flights from Gatwick and Southend. Their aircraft also occasionally visited Birmingham operating charter flights to Ostend and Jersey. By the end of September, though, work was becoming scarce, and the Manchester to Ostend services stopped, and the airline was now relying on ad-hoc charter work - a Manchester to Nantes charter flight was operated by the Viking on October 9, and a Jersey to Manchester charter on October 16. The Viking also visited Heathrow operating flights for other airlines, and on November 12 operated a Gatwick to Amsterdam service.

Prior to this, in August Air Condor had hoped to operate a DC4 (G-AREL) on its inclusive tour charters, but the deal for the lease of this aircraft fell through, and the aircraft was never flown by Air Condor. Later, another DC4 (G-APID) was sprayed in Air Condor's colours at Heathrow Airport in November 1960, but again this aircraft was never delivered to the airline. With the winter approaching, Air Condor's services were run down - one of the airline's last charter flights was made by the Viking on November 19 when G-AKTU flew from Amsterdam to Gatwick, before positioning back to Southend later the same day. This was followed by the company appointing a receiver on December 4 1960, and subsequently being placed into liquidation.

Fleet:

Bristol 170 Freighter Mk.1A	G-AGPV	c/n 12730	dd. 3.60,	ISOC.
Bristol 170 Freighter Mk.21	G-AHJD	c/n 12736	lsd. 4.60,	ret. 8.60, leased from North South Airlines.
Vickers 634 Viking	G-AKTU	c/n 211	lsd. 5.60,	ISOC , leased from Keegan Aviation.
Douglas C54A Skymaster	G-AREL	c/n 10563	rd. 8.60,	never entered service.

AIR CONTRACTORS

Air Contractors Ltd. Bases: Woodley, Blackbushe and later Bovingdon Aerodromes.

Air Contractors was an early post-war charter operator founded on May 15 1946 by Group Captain P.H.Maxwell. Based at Woodley, this company started passenger and freight ad-hoc charter operations with the prototype Miles Aerovan on June 2 1946. The company held a contract with the Miles Aircraft Company for the transport of unfinished Miles Messengers from the Miles' Newtownards factory to their headquarters at Woodley Aerodrome. As business built up Air Contractors acquired further Miles Aerovans, and by March 1947 seven aircraft were in service flying passenger and freight charters to many parts of Europe carrying numerous different types of cargoes. The first of three Dakotas entered service in January 1947, and the introduction of this type of aircraft led to the airline's move to Blackbushe Aerodrome. From Blackbushe the Dakotas, and later a Bristol Freighter, were used on numerous freight charters to North Africa, the Middle East and Europe. Cargoes included fruit, vegetables, racehorses and cattle, and in addition Air Contractors flew forces leave contracts to the Middle East on behalf of the Ministry of Defence.

The Aerovans were often used for passenger charters, and during a strike of French railway employees in June 1947, these aircraft bought back over one hundred and fifty stranded British tourists from Paris to Blackbushe on June 14 and 15 - each Aerovan was fitted out with six seats, and the company made almost thirty return flights bringing back these holidaymakers. During 1947 the Dakotas made many flights to North Africa with cattle, and at the end of the year the airline started flying refugees between India and Pakistan. All three Dakotas were used on this task, and when the three aircraft returned to the United Kingdom in November 1947 they had carried a total of six thousand refugees over a total of 105,000 miles. On one flight, one of the Dakotas had carried a remarkable total of sixty five passengers, but not all flights were quite so heavily loaded!

Shortly after the Dakota's return from India, the airline again moved its base - this time to Bovindon early in 1948. During the spring several fruit and vegetable charters were operated from towns on the Continent to Bovingdon, from where the produce continued its journey by road to markets in London and the Midlands. Several livestock charters were carried out to Florence, and in April and June a Dakota made cattle carrying flights to Nairobi. By this time only a few Aerovans were left in service, and those still in use were mainly flown on short haul freight charters, but one aircraft did venture as far as Zurich in June carrying an aero engine, and a similar engine was brought back on the return journey. When the Berlin Airlift started in August 1948, Air Contractors was one of the first British airlines to take part, and on August 4 all three Dakotas were positioned to Fassberg to start flying sorties on the airlift. The aircraft were moved to Lubeck on August 28 and flying continued from here until November 10 1948 when, like other Dakota operators, Air Contractors' aircraft were withdrawn from the airlift, and flown back to the United Kingdom. During its three months on the airlift, Air Contractors had flown 386 sorties, carrying a total of 1,376 tons of freight in one thousand and sixty six flying hours. Of the three Dakotas, G-AIWC flew the most sorties with one hundred and seventy two to its credit, followed closely by G-AIWE with one hundred and sixty one sorties.

However, the end of the airlift was a time of depression for Air Contractors, and within a month of the aircraft returning to Bovingdon, the entire fleet was sold to Skyways, and the aircraft migrated south to Dunsfold during December. Air Contactors still had a handful of serviceable Aerovans, and attempts were made to continue flying with these aircraft, but work was scarce and by the end of January 1949 all flying had ceased, and Air Contractors closed down.

Fleet:

Miles M.57 Aerovan 1	G-AGOZ	c/n	4700	dd. 6.46,	sold 1.47.
Miles M.57 Aerovan 3	G-AHXH	c/n	6382	dd. 3.47,	sold 6.48, converted to series 4.
	G-AHTX	c/n	6380	dd. 6.46,	ISOC.
	G-AIDI	c/n	6383	dd. 9.46,	ISOC.
	G-AIHK	c/n	6384	dd.10.46,	ISOC.
	G-AIHL	c/n	6385	dd.10.46,	written off Copenhagen 29.10.46.
Miles M.57 Aerovan 4	G-AIKV	c/n	6389	dd.12.46,	written off Cherbourg 12.1.47.
	G-AISI	c/n	6397	dd. 3.47,	sold 5.48.
Bristol 170 Freighter Mk.1	G-AHJD	c/n	12736	dd. 7.47,	sold 10.48.
Douglas C47A Dakota	G-AIWC	c/n	13474	dd. 4.47,	sold 12.48.
	G-AIWD	c/n	13475	dd. 7.47,	sold 12.48.
	G-AIWE	c/n	13479	dd. 1.47,	sold 12.48.
Miles M.65 Gemini 1A	G-AISO	c/n	6326	dd. 4.47,	sold 12.47.

AIR COURIERS

Air Couriers (Transport) Ltd.

Bases: Croydon, Biggin Hill and later Gatwick Airport.

Air Couriers was formed in 1938 as an aircraft engineering and operating company by Mr.F.W.Griffith. Based at Croydon, the company undertook aircraft maintenance and overhaul, but operations were suspended when war broke out in 1939. However, once the war had ended, Air Couriers resumed its maintenance activities at Croydon, and also formed a charter division known as Air Couriers (Transport) Ltd. This new company started ad-hoc passenger and freight charter operations from Croydon early in 1948, and its directors were Messrs G.G.Glanville F.W.Grffith and C.P.L.Godsal. The company flew a varied fleet consisting of Geminis, Proctors and Rapides on air taxi services, freight charters, pleasure flying and aerial photography. Over the next fifteen years a wide variety of services were offered, including a large amount of work undertaken on behalf of the Press, and for film and television. Air Couriers became responsible for the conversion of many Rapides and Ansons from their military configuration to civilian standards, and although many of these aircraft were sold, Air Couriers did retain some of the Rapides for use on charter work by their associate - Air Couriers (Transport) Ltd.

The most common of all Air Courier's Rapides was G-AKNY which travelled over most parts of the British Isles and many parts of Europe during its seven years of service with Air Couriers (Transport). During 1959 three Rapides were in service, and the aircraft were regular visitors to Manchester and Blackpool during the summer and numerous flights were made between Blackbushe and Blackpool on Sundays. However, Croydon Airport closed in September 1959 and Air Couriers moved its base to Biggin Hill where it acquired hanger space, and continued its maintenance activities from here. During 1960 Air Couriers undertook the maintenance of several large types of aircraft including the Vikings of Maitland Drewery, and often the DC4s of World Wide Airways would receive Air Couriers' expert attention. In addition many newly imported American Piper aircraft were anglicised by Air Couriers, and at the end of the 1960 summer season the fleet of Rapides was sold. This temporarily marked the cessation of any charter work carried out by Air Couriers, but later, in 1961 another Rapide was acquired, and this was again used on charter work, aerial photography and pleasure flying from Biggin Hill. But the days of this Rapide, too were numbered and in 1963 this aircraft was retired, and replaced by one of the new breed of American Piper Apaches. By this time Air Couriers had its head office at Gatwick Airport where it undertook the maintenance of many airlines' aircraft - the premises at Biggin Hill were retained to undertake light aircraft maintenance only.

The Piper Apache was regularly used by Air Couriers (Transport) for training flights from Gatwick Airport, and during its service with the company it operated several charter flights to continental destinations, and also many airports in the British Isles including Bembridge, Cardiff, Hawarden and Manchester. This activity continued until July 1967 when Air Couriers and Air Couriers (Transport) were both bought out by the Gatwick based Transglobe Airways - at the time of this take-over, both Mr.F.W.Griffith and his son, Mr.E.S.Griffith, were still managing the company. After the take-over, both men resigned, and the name of Air Couriers disappeared after almost thirty years of service to the aviation industry.

Fleet:

Percival Proctor 1	G-AHVB	c/n	K.292	rd. 3.48,	sold 9.50.
	G-AHVC	c/n	K.324	rd. 3.48,	wfu. 5.51.
Percival Proctor 3	G-AJWN	c/n	H.195	ca.10.47,	sold 1.50.
Short Scion Junior	G-AEZF	c/n	PA1008	dd. 8.46,	sold 8.47.
Messerschmidt BF108D-1	G-AKZY	c/n	3059	dd. .49,	sold 1.50, never entered service.
DH89A Dragon Rapide	G-AKNN	c/n	6598	dd. 6.51,	sold 6.54.
	G-AKNY	c/n	6470	dd. 8.53,	sold 10.60.
	G-AKSE	c/n	6870	dd. 8.61,	sold 4.63.
	G-AKTD	c/n	6791	ca. 4.48,	sold 5.51.
	G-ALBI	c/n	6525	dd. 6.55,	sold 8.60.
	G-APBM	c/n	6748	ca. 9.57,	sold 8.60.
Miles M.65 Gemini 1A	G-AKEG	c/n	6299	dd.11.55,	sold .56.
	G-AKFY	c/n	6503	rd. 1.54,	sold .60.
	G-AKHV	c/n	6523	rd.11.56,	sold .58.
Reid & Sigrist RS4 Bobsleigh	G-AGOS	c/n	3	dd. 3.56,	sold 3.58, converted to RS4 Desford Trainer January 1958.
Airspeed Oxford	G-AHGU	c/n	3277	dd.10.56,	sold 7.60.
Piper PA23 Apache 160	G-ARJR	c/n	23-1966	dd. 4.61,	ISOC.

AIRCRAFT AND ENGINEERING

Aircraft and Engineering Services Ltd.

Base: Croydon Airport.

Aircraft and Engineering Services was registered as a charter operator and a maintenance company in July 1946. Based at Croydon Airport the company started air charter operations in September 1946 with one Dragon Rapide which was replaced in March 1947 by a Percival Proctor. At Croydon the company undertook the overhaul and conversion of many Royal Air Force aircraft to civilian standards, and these aircraft included Tiger Moths, Proctors and Rapides. The company concentrated on this side of the business, and although a small amount of charter work was undertaken and the aircraft were occasionally hired out for private use, the charter side of the business had closed down by the spring of 1950. During its three years of charter flying, Aircraft and Engineering Services had used Tiger Moths, Rapides and Proctors for pleasure flying and flying training from Croydon Airport.

Fleet:

Percival Proctor 2A	G-AHNB	c/n	H.186	dd. 3.47,	wfu. 10.48.
DH89 Dragon Rapide	G-ACPP	c/n	6254	rd. 4.48,	sold 6.48.
DH89A Dragon Rapide	G-AFHY	c/n	6417	dd. 9.46,	sold 3.47.
	G-AJMY	c/n	6511	ca. 2.48,	sold 3.48.
	G-AKVU	c/n	6476	ca. 3.49,	ISOC.

AIRCRUISE

Aircruise (Operating) Ltd. Bases: Cranfield and Leavesden Airports.

Formed on August 28 1963, Aircruise started air taxi operations from its base at Cranfield with a Piper Aztec and Beechcraft Travelair. The aircraft owned by the company were mainly registered in the name of one of the airline's owners - Mrs.J.M.Breeden. Two de Havilland Doves were acquired in April 1964, and this coincided with the company's move to a new base at Leavesden Airport. Further Doves were rapidly bought, and they were mainly used on executive charters from Leavesden. In addition, Aircruise held contracts with several British companies for the provision of air charter facilities, and the Doves became familiar sights at airports, both at home and abroad. Many of the aircraft flew services on behalf of Rolls Royce's aero engine division, flying executives between the company's factories at Leavesden, Filton and Hucknall. One of the Doves was based at Southend during 1964 to fly executives of the Ford Motor Company to Birmingham - sometimes the Dove would fly twice daily in each direction, and if loads were too heavy for one Dove, the aircraft would be assisted by a second Dove or a Piper Aztec. The Doves were also leased out to other carriers - one aircraft spent nine months with South Coast Air Services operating charters and pleasure flights from Shoreham, while two other aircraft were to have been leased to Algenova for use on their Pisa to Elba shuttle service, but these aircraft never reached Italy. However, one Dove was severely damaged in a force landing near Droitwich in April 1965 and after this the company's operations were run down, although a number of Cessna and Piper twins continued to be flown by the airline on air taxi work from Leavesden for another four years. Flying training was also undertaken, but when the company finally closed down early in 1970, only one Cessna 172 was left in service.

Fleet:

DH104 Dove 1	G-AICY	c/n	04019	dd. 6.64,	sold 12.64.
	G-AJDP	c/n	04028	dd. 7.64,	sold 2.66.
DH104 Dove 2	G-AJLW	c/n	04033	dd. 4.64,	written off Droitwich 26.4.65.
DH104 Dove 5	G-ASDD	c/n	04452	dd. 8.64,	sold 7.65.
	G-ASMX	c/n	04486	dd. 4.64,	sold 5.65.
Piper PA27 Aztec 250A	G-ARBR	c/n	27-170	dd. 8.63,	sold 5.65.
Piper PA27 Aztec 250B	G-ATCM	c/n	27-2834	dd. 3.68,	written off Brest, France 26.6.68.
Piper PA30 Twin Comanche	G-ASLD	c/n	30-58	dd. 8.65,	sold 12.66.
Cessna F.172G	G-ATKU	c/n	0232	dd. 3.66,	ISOC.
Cessna 337B Super Skymaster	G-AVIX	c/n	0544	dd. 4.67,	sold 9.69.
Cessna 310F	G-AROK	c/n	0148	dd. .64,	sold 7.65.
Cessna 310J	G-ATCS	c/n	0145	dd. .66,	sold 4.70.
Beech 95 Travelair	G-APUB	c/n	TD240	dd. 8.63,	sold 10.63.

AIR ENTERPRISES

Air Enterprises Ltd. Bases: Croydon and Renfrew Airports.

Air Enterprises was founded as an airline company by Henry Carr, Arthur Carr, Francis Carr and Thomas Hudson on April 25 1947. Based at Croydon the airline started passenger and freight ad-hoc charter operations in May 1947 with a Percival Proctor and an Airspeed Consul. By the end of 1947 a Dragon Rapide and a further Consul had joined the company, and a base was set up at Renfrew Airport, Glasgow. From Renfrew one aircraft was used on ad-hoc charter flights to many of Scotland's scattered airfields, and generally one Dragon Rapide was used on these services. Between late 1947 and 1949 Air Enterprises operated air services on behalf of the United Nations Commission in Israel with its Consuls.

Air Enterprises was one of the first airlines to apply to operate passenger services under a B.E.A. Associate Agreement. During the spring of 1948 the company applied to open services to Cowes from Southampton, Gatwick and Croydon, with an extension of the Croydon to Cowes service onto Sandown. These services were opened in July 1948 using a fleet of newly acquired Rapides, and were mainly flown at weekends, before being suspended for the winter in September 1948. For the 1949 summer season the airline again operated a Croydon to Cowes and Sandown service, but Air Enterprises also received approval to operate scheduled services from London to Birmingham and Manchester, but whether these latter services were ever operated is not known. The services to the Isle of Wight were again suspended at the end of the summer, and during the winter months the company undertook many charter flights, and passenger charters were often operated to continental airports. During 1950 Air Enterprises continued to operate numerous scheduled services to the Isle of Wight with a fleet of four Rapides. An Anson was acquired in January 1950, but this aircraft never entered service with Air Enterprises, and was sold a year later to another airline.

By 1951, in addition to the Croydon to Sandown service, Air Enterprises also flew scheduled services to Sandown from Heathrow, Northolt and Gatwick. Services were also opened from Shoreham to Jersey and Sandown - the service to Sandown was flown via Portsmouth. These services were flown from June until September, and Air Enterprises' aircraft were also often used for charter flights from Croydon to various race meetings around the country. One of the airline's founders, Mr.Francis Carr, was chairman of the company at this time, but the position of managing director had been taken by Captain John Dada, who had previously been on the board of directors of Morton Air Services before joining Air Enterprises in 1948. Captain W.G.Poulter became chief pilot and operations manager of the airline.

The services to the Isle of Wight and Jersey continued during 1952, and in October Air Enterprises took delivery of a Percival Prince which it operated on behalf of the Standard Motor Company. However, after little more than three months of flying the Prince was extensively damaged when it force landed just short of Manston's runway on February 6 1953. For the 1953 season, the company's fleet was run down to four Airspeed Consuls, and at the end of the season the company closed down, and sold its fleet of aircraft.

Fleet:

DH89A Dragon Rapide	G-AFMJ	c/n	6436	dd.12.47,	sold 4.53,	'The Shanklin Flyer'.
	G-AKNX	c/n	6629	dd. 2.48,	sold 1.49.	
	G-AKNY	c/n	6470	dd. 3.48,	sold 1.49.	
	G-AKNZ	c/n	6550	dd. 3.48,	sold 12.48.	
	G-AKOA	c/n	6618	dd. 3.48,	sold 1.49.	

DH89A Dragon Rapide	G-AKOB	c/n	6492	dd. 3.48, sold 10.52,	'The Sandown Flyer'.
	G-AKRS	c/n	6952	dd. 4.48, sold 4.53.	
	G-ALWY	c/n	6741	dd. 1.50, written off	Isle of Islay 19.4.52, 'The Ventnor Flyer'.
Percival Proctor 1	G-AIHG	c/n	H.4	rd. 5.47, sold 6.50.	
Airspeed AS.65 Consul	G-AGVY	c/n	3204	dd.12.48, written off	Lebanon 11.2.49.
	G-AIKT	c/n	4356	dd. 6.51, ISOC.	
	G-AIOT	c/n	4330	dd.12.48, ISOC.	
	G-AIOV	c/n	4361	dd.12.48, ISOC.	
	G-AJLJ	c/n	5128	dd. 5.47, wfu. 3.50.	
	G-AJLK	c/n	5129	dd. 6.47, sold 1.51.	
	G-AJNE	c/n	5139	dd.10.48, ISOC.	
Avro 652A Anson	G-ALWX	ex.	EG647	dd. 1.50, sold 1.51,	never entered service.
Percival P.50 Prince	G-AMPR	c/n	P50/48	dd.10.52, ISOC,	operated on behalf of the Standard Motor Company.

AIR FERRY

Air Ferry Ltd. Base: Manston Aerodrome.

Wing Commander Hugh Kennard resigned from the board of directors of Silver City Airways in November 1960 after being on the board of that company since 1953, in which year Kennard's Air Kruise was bought out by the British Aviation Services group. Air Kruise had, at one time, been carrying more inclusive tour charter passengers than almost all the other tour operators put together. In May 1961, therefore, Wing Commander Kennard announced that he would shortly be returning to the air transport business, and in July 1962 it was announced that his new company would be called Air Ferry.

Air Ferry applied, in conjunction with Leroy Tours, for numerous licences to operate inclusive tour charters from Manston Airport (the airline's base) with a fleet of Douglas DC4 and DC6 aircraft. While the position of managing director of Air Ferry was taken by Wing Commander Kennard, Mr.Lewis A.Leroy (managing director and owner of Leroy Tours) became Air Ferry's chairman. Air Ferry's first aircraft - two Vickers Vikings and one DC4 - arrived at Manston in February 1963, and after a period of crew training and a series of route proving flights, Air Ferry began operations on March 30 1963. During the summer months inclusive tour charter flights were operated to more than twenty European destinations including Verona, Rimini, Venice, Ostend Palma, Malaga, Perpignan, Barcelona, Pisa, Basle, Copenhagen, Dijon, Geneva, Naples, Luxembourg, Rotterdam, and the Canary Islands. In addition, the company flew services from Manston to Le Touquet and Ostend on a scheduled basis, and although a licence was also held for a scheduled service between Manston and Verona, this route was not opened. During 1963 Air Ferry applied to open vehicle ferry services from Manston to Dijon, Calais and Toulouse, but these applications were turned down by the Air Transport Licensing Board. However, Air Ferry did receive permission to operate a scheduled service between Belfast and Le Touquet using DC4, DC6 or Viking aircraft, but this service was never started. Air Ferry's first summer season came to an end on October 20 1963, and by this time over one hundred and twenty thousand passengers had been carried on its services during seven months of operations.

For the 1964 season Air Ferry had five Vikings and three DC4s in service, and these aircraft were employed on similar services to those operated during 1963. However, at the end of the 1964 summer season, Air Ferry was bought out by the Air Holdings group in October 1964, and although Air Ferry's operations continued unchanged, Wing Commander Kennard left the airline to form Invicta Airways, another Manston based inclusive tour operator. By 1965 Air Ferry was operating regular scheduled vehicle ferry services between Manston and Le Touquet using Bristol Freighters, and later services were introduced serving Calais, Ostend and Rotterdam for the carriage of cars, passengers and cargo. Air Ferry had hoped to open its Belfast to Le Touquet route using Carvairs, but this service was never started, and plans were announced for the operation of vehicle ferry services from Glasgow and Leeds to the Continent. In December 1965 two of British United's Carvairs were leased to Air Ferry to carry essential supplies into Zambia after the Rhodesian declaration of U.D.I. in November. The first Carvair (G-ASKG) left Southend on December 28, and returned on March 11 1966, while the second aircraft remained in Zambia until its return on June 2 1966. These aircraft flew from Dar-es-Salaam to Lusaka, and Air Ferry also sent two of its DC4s (G-ASFY and G-ASOG) to Lusaka for these services. This work helped to keep the DC4s busy during the normally slack winter months for tour operators.

The DC4s were also operated on some of British European Airways' and Lufthansa's International scheduled freight services from the United Kingdom to the Continent. From Manchester, one of Air Ferry's DC4s flew services to Dusseldorf and Zurich and from London the DC4s flew to Dusseldorf, Frankfurt and Zurich - however, one of the DC4s crashed on approach to Frankfurt while operating one of these services on January 21 1967 killing the crew. After this, B.E.A. took over most of these services using Vanguard and Argosy aircraft, and during the summer of 1967 Air Ferry's surviving DC4s returned to their normal work of flying tourists to their continental holiday resorts from Manston, Manchester, Gatwick and Newcastle on behalf of Lyons Tours and Leroy Tours. The airline had taken over British United's two Douglas DC6s in December 1965, and these aircraft were also used on tour services, although they were mainly used for long range freight charter work. 1967 was not a good year for Air Ferry - after losing one DC4 in the fatal accident at Frankfurt, a second DC4 was lost at the height of the 1967 season when it crashed into high ground while decending into Perpignan on a tour flight from Manston. Tragically, everyone on board was killed, and this left Air Ferry with just one DC4 to carry on its services, resulting in the lease of another aircraft from Lloyd International.

By 1968 most inclusive tour charter operators had replaced their older piston-engined aircraft with more modern turbine equipment, and consequently Air Ferry negotiated the lease of two Channel Airways Viscounts to operate its services during the summer of 1968. Both aircraft were sprayed up in full Air Ferry colours and during the summer these aircraft flew many thousands of passengers from Manston, Gatwick, Newcastle and Manchester on tour services on behalf of several British tour operators including Lyons Tours and Leroy Tours. In addition a nightime service was also flown from Bristol to Perpignan - this service, operated on behalf of Lyons Tours, was flown once every ten days. Several ad-hoc charter flights were also operated from Southend to the Continent, and the Viscounts often visited many other British airports operating a variety of passenger charters. However, early in the season Air Holdings announced that Air Ferry would be closed down at the end of the season, and all services transferred to British United Airways. Consequently, on October 31 1968 the airline operated its last revenue earning flight - the Viscounts were returned to Channel Airways and the other aircraft were sold.

Fleet:

Vickers 498 Viking	G-AHOW	c/n	124	dd. 4.64, sold 11.66.	
Vickers 610 Viking	G-AIVD	c/n	217	dd. 2.63, wfu. 4.65.	
	G-AIVF	c/n	219	dd. 2.63, sold 4.66.	
	G-AJBX	c/n	249	dd. 4.64, wfu. 5.65.	
Vickers 621 Viking	G-AOCH	c/n	150	dd. 8.63, sold 4.66.	
Douglas C54A Skymaster	G-APYK	c/n	10279	dd. 2.63, written off	Pyrenees 3.6.67.
	G-ASFY	c/n	10335	dd. 3.63, ISOC.	
	G-ASOG	c/n	10359	dd. 1.64, written off	Frankfurt 21.1.67.
Douglas C54B Skymaster	G-ARWI	c/n	18349	lsd. 6.67, ret. 10.67,	leased from Lloyd International.
Bristol 170 Freighter Mk.31	G-AMLL	c/n	13074	lsd. 3.64, ret. 4.65,	leased from Hondley Page.
Bristol 170 Freighter Mk.32	G-ANVR	c/n	13251	lsd. 1.65, ret. 3.65,	leased from B.U.A.F.

G-AGJV is depicted here about to leap into the air on another joy riding sortie at the 1974 Greenham Common Air Display.

Tony Merton Jones

G-BCDN was the third Friendship to be delivered to Air Anglia, and it is caught here one summer's day at Aberdeen during 1974.

A J Large

Alidair's Viscount G-AZNH had the distinction of operating a special flight on July 29 1975 between Northolt and Le Bourget, to celebrate the twenty-fifth anniversary of the first flight by a turboprop aircraft between the cities of London and Paris.

British Aircraft Corporation

Although G-AGPV was the prototype Bristol Freighter, first flying at Filton on December 2 1945, it was still flying holidaymakers to the Continent in 1960 in the colours of Air Condor.
A J Jackson

G-AISI was one of Air Contractors' many Miles Aerovans.
A J Jackson

Air Enterprises' Consuls survived longer than most others of the type on the British register, flying scheduled services and U.N. flights in Israel.
A J Jackson

G-AFMJ Air Enterprises Dragon Rapide A.J.Jackson

G-AKNY Air Couriers Dragon Rapide Ian MacFarlane

G-AZXG George House Aviation Aztec Tony Merton Jones

G-.... Airgo Aztec Stan Davidson

G-ASFY Air Ferry Douglas DC-4 Ian MacFarlane

G-BADF Airgo Seneca Stan Davidson

Douglas DC6A	G-APNO	c/n	45531	dd.12.65,	ISOC.
	G-APNP	c/n	45532	dd.12.65,	ISOC.
Vickers 812 Viscount	G-AVHE	c/n	363	lsd. 1.68,	ret. 10.68, leased from Channel Airways.
	G-AVNJ	c/n	361	lsd. 1.68,	ret. 10.68, leased from Channel Airways.
ATL.98 Carvair	G-APNH	c/n	11	lsd.12.65,	ret. 6.66, leased from B.U.A.F.
	G-ASKG	c/n	10	lsd.12.65,	ret. 3.66, leased from B.U.A.F.

AIR FREIGHT

Air Freight Ltd. — Bases: Gatwick, Bovingdon and Luton Airports.

Air Freight was formed late in 1946 as a freight operator, and registered as an airline company on February 28 1947 by its three founders - Messrs H.K.Hughes, E.D.Nicol and F.A.Dickson. By this time the company was operating a fleet of two Halifaxes on long distance freight charter flights to the Far East and North Africa, and also on many flights to the Continent to collect fresh fruit and vegetables which the aircraft then brought back to Bovingdon for markets in London and the Midlands. Air Freight mainly used Bovingdon for all its services, although visits were often made to Gatwick and Luton airports. However, early in 1948 the airline was bought out by Mr.Harold Bamberg, founder of the newly formed Eagle Aviation, and the operations of Air Freight were integrated with those of Eagle. Thus the name of Air Freight temporarily died, but in September 1952 the name was revived by Eagle when three Avro Yorks were bought and placed into service under the name of Air Freight. However, this revival was short lived because the company was bought out by the Lancashire Aircraft Corporation in July 1953, and the Yorks passed into service with Skyways.

Fleet:

Percival Proctor 5	G-AIAG	c/n	Ae.88	dd. .47,	ISOT.
Handley Page 61 Halifax 8	G-AJBL	c/n	1349	dd. 1.47,	ISOT.
	G-AJBK	c/n	1337	dd. 1.47,	sold 10.47.
Handley Page 61 Halifax 6	G-AKJI	ex.	RG695	dd.10.47,	never entered service.
	G-AKJJ	ex.	RG698	dd.10.47,	never entered service.
Avro 685 York 1	G-AGNP	c/n	1218	dd. 9.52,	while under Eagle ownership.
	G-AGNT	c/n	1221	dd. 9.52,	while under Eagle ownership.
	G-AGNU	c/n	1222	dd. 9.52,	while under Eagle ownership.

AIRGO

Airgo Ltd.
Scotcourt Executive Air Ltd.
Glasgow Air Taxi Services Ltd.

Bases: Glasgow (Abbotsinch) and outstations at Aberdeen, Norwich and Heathrow Airports.

Glasgow Air Taxi Services was formed on July 20 1972 by Mr.Ken Stuart. Air taxi operations began in September with a leased Cherokee Six, which was replaced later that month with a new Piper Seneca. Trading under the name of Airgo, this abbotsinch based charter operator expanded very rapidly. By early 1973 a number of Aztecs were operated, and much work was carried on from Aberdeen where Airgo set up an outstation, and several aircraft were then based there to operate flights on behalf of several oil companies. The company also flew an Aztec on behalf of Site Aviation,but this arrangement came to an end in August 1973 when Site employed its own crews to fly its aircraft. Throughout 1973 and into 1974, Airgo acquired Aztecs, Navajos and a second Seneca, and also leased other aircraft; an outstation was set up at Heathrow and an Aztec based there, and at Norwich, Airgo based a Navajo Chieftan which was used for flights to Aberdeen, Amsterdam and across the North Sea to Stavanger and Bergen on behalf of oil companies. Gradually, Glasgow became less important in the company's network, and Aberdeen grew in importance. In addition to looking after its own aircraft at Aberdeen, Airgo handled aircraft for other carriers including Air Bridge Carriers and British Midland Airways and at Glasgow the airline handled Aer Arann.

However, these operations were not without their problems. Airgo was operating a large fleet of sophisticated twin engined aircraft into numerous airfield and airstrips throughout Scotland, and several aircraft were damaged in landing and take-off accidents at some of the poorer equipped airfields. Expansion, though, continued and during 1973 a Hovercraft was purchased to operate a ferry service from Dalcross Airport, Inverness to Niggs Bay on the Cromarty Forth. In January 1974, another Scottish air taxi operator, Scotcourt Executive Air, was taken over, and this added a further three aircraft aircraft to Airgo's ever increasing fleet. However, in October 1974 the bubble burst when Airgo was bought out by the Edinburgh based Lowland Aero Services. Lowland's, who had been operating a varied fleet of Cessna training aircraft, immediately rationalised Airgo's operations. The Enstrom helicopter, Navajos, Senecas and several Aztecs sold, and half the company's staff was made redundant. The bases at Heathrow and Norwich were closed, and the operations base moved to Edinburgh, with a second base at Aberdeen Airport. A fleet of Aztecs was retained to operate air taxi services from Aberdeen and Edinburgh, and these aircraft flew most of their services within Scotland, although occasional trips were made to Amsterdam, Stavanger, London, Manchester and even to the races at Newbury. However, early in August 1975 most of Airgo's operations were closed down and many aircraft sold - a Gill Aviation Aztec (G-BBCC) was retained until early September operating air taxi services, but once this aircraft had returned to Gill, no further flying was undertaken.

Fleet:

Piper PA32 Cherokee Six	G-ATTY	c/n 32-460	lsd. 9.72,	ret. 9.72.
Piper PA34 Seneca 200	G-BADF	c/n 34-50257	dd. 9.72,	sold 4.75.
	G-BBFF	c/n 34-50076	dd. 9.73,	damaged Colonsay 25.5.74.
Piper PA27 Aztec 250C	G-BAED	c/n 27-3864	dd. 8.73,	ISOC.
Piper PA27 Aztec 250D	G-AYEM	c/n 27-4411	dd.11.72,	ISOC.
	G-AYZC	c/n 27-4570	dd.10.73,	ISOC, 'Airgo Aberdeen'.
	G-BAUA	c/n 27-4048	dd.12.73,	sold 7.74.
	G-BBCC	c/n 27-4317	lsd. .75,	ret. 9.75, leased from Gill Aviation.
	G-BBSR	c/n 27-4394	dd. 1.74,	sold 5.75.
Piper PA27 Aztec 250E	G-AZFE	c/n 27-4577	lsd. .73,	ret. 8.73, leased from Site Aviation.
	G-BATX	c/n 27-4832	dd. 4.74,	ISOC.
	G-BBCW	c/n 27-4805	dd. 1.74,	ISOC, 'Airgo London'.
	G-BBMJ	c/n 27-5150	lsd.12.73,	ret. 12.73, leased from Webster Aviation.
Piper PA31 Navajo 300	G-AYVM	c/n 31-516	dd. 1.74,	sold 9.74.
	G-AZBP	c/n 31-551	dd. 2.74,	sold 6.74.
Piper PA31 Navajo 350	G-BBNT	c/n 31-05107	dd. 4.74,	ISOC, 'Airgo Norwich'.
Bell 206A Jetranger	G-AYCM	c/n 529	lsd. 2.73,	ret. .73, leased from Fras-Air.
Enstrom F28A	G-BBPO	c/n 176	dd. 3.74,	sold 1.75.
Cessna 310Q	G-AZYM	c/n 0507	dd. 1.74,	sold 3.74.
Cessna F172M	G-BAXY	c/n 0905	dd. 1.74,	sold 1.75.

AIR GREGORY

Air Gregory Ltd.
Gregory Air Services Ltd.
Gregory Air Taxis Ltd.

Bases: Main base Denham, but the HS.125s were based at Luton, and DC3 at Newcastle.

Gregory Air Taxis was formed early in 1963 by Mr.Ken Gregory - an ex-racing driver - to provide a twenty four hour service to businessmen requiring an air taxi service. Initially, a single Comanche was flown from the company's base at Denham airfield on air taxi flights, but very quickly the demand for this type of service was so great that Piper Aztecs and Twin Comanches were soon added to the fleet. Gregory also became a distributor for the Swiss Pilatus Porter aircraft, and an example of this aeroplane was flown by Gregorys for just over six months on demonstrations and charters from Denham. During 1963 a Denham based maintenance organisation, Marshall Dunn Aero Services, was taken over by Gregorys and on February 24 1964 a new company named Gregory Air Engineering was registered by Mrs.Anemone C.Gregory and Mr.Kenneth A.Gregory to take over the services provided by Marshall Dunn Aero Services.

1964 was a year of expansion for Gregory Air Taxi - by May two Aztecs and two Twin Comanches were in service operating air taxi flights to many parts of the British Isles and Europe. With no customs facilities available at Denham, many of Gregory's international flights staged through Heathrow or Gatwick, and Gregory's aircraft became familiar sights at many of Britain's major airfields. Gregory's high standard of service soon became known and the company became one of the foremost specialists in this type of aviation. This all led to further expansion, and on March 9 1964 another company, Gregory Air Taxis (Liverpool) Ltd., was registered and on April 1 this company started air taxi operations from Liverpool Airport with one Piper Aztec (G-ASHH). This operation lasted for two years, and in 1966 the base was closed and the aircraft returned to Denham. From Liverpool the Aztec had flown a variety of domestic and international charters, and the initial success of this operation led Gregory to open a base in the North East of England. The airfield chosen for this base was Sunderland Airport, and on May 1 1964 one Twin Comanche (G-ASYK) started flying air taxi services from Sunderland.

The base at Sunderland remained open for a year, and on April 3 1965 Gregory's first 'large' aircraft, a Dakota, arrived at Newcastle Airport to start flying charters from the North East of England. This was followed on April 4 by the transfer of Sunderland's Twin Comanche to Newcastle, and its air taxi operations were now conducted from here. During the spring further bases were opened at Gatwick and Plymouth, and at each airport Gregory based one aircraft - the arrival of the Dakota had seen the operating name of Air Gregory being adopted. The Dakota had been acquired to operate inclusive tour charters from Newcastle to Beauvais, Luxembourg and Ostend, and although many services were flown from Newcastle by this aircraft, most of its time was spent on ad-hoc charter work and on lease to other airline companies. The introduction of the Dakota in April 1965 was followed by the introduction of an HS125 executive jet in December 1965. To operate this aircraft, and the Dakota, Gregory formed a new company named Gregory Air Services which became responsible for the operations of all aircraft above 12,500 lb all up weight. Air Gregory was the first air taxi company to introduce an executive jet, but due to the lack of a hard runway at Denham the 125 was based at Luton Airport. From Luton this aircraft, operated in association with Air Hanson, flew numerous international charter flights to many European capital cities, and during its first year of operations with Gregory the utilisation of this aircraft was far higher than had been expected.

During 1966 Gregory negotiated with several tour firms for inclusive tour charter flight contracts for the 1967 summer season, but no contracts were placed with Gregory. Had these negotiations been successful, Gregory would have either have operated Comets from Manchester and Liverpool, or Viscounts from Bristol and Cardiff. However, Gregory continued executive charter operations normally until December 1968 when the Bristol Street motor group bought out the Air Gregory group of companies, and in 1969 it was decided to sell the HS125 fleet due to non-profitability. This, in turn, led to Ken Gregory's resignation from the board of directors, and subsequently the air taxi side of the business was run down, and Air Gregory concentrated on flying training, both in fixed wing aircraft and in helicopters. At the time of the sale of the HS125 the aircraft had reached an average annual utilisation of 600 hours, with up to ninety hours being achieved during the peak summer months and about twenty three hours a month during the normally slack winter period. Mr.Gregory left Air Gregory to concentrate on his publishing business. Air Gregory's most profitable operation at this time was concerned with the flying training and aircraft sales side of the business, and it is this which the Bristol Street Group now concentrated on. Air Gregory also flew a large fleet of light and medium sized helicopters from Denham, and in addition to air taxi work with these aircraft, the company also carried out a large amount of aerial photography and work for the Press, film and television companies. The sale of new and many second hand light aircraft was also one of Air Gregory's major occupations, and by 1971 the air taxi fleet had been run down completely.

Early in 1971, Air Gregory was acquired by the Medminster Group, which also runs Kentair Charters and the Surrey and Kent Flying Club. Subsequently Kentair transferred one of its Aztecs to Air Gregory in October 1972 and flying in Air Gregory's colours from Denham, this aircraft reopened Air Gregory's air taxi services. Another Aztec was added two years later in November 1974, and these aircraft now operate a wide variety of air taxi flights within the British Isles and Europe. On March 10 1975 the fixed wing part of the flying school was sold to Mr.T.W.Howard and the name of the school changed to the Denham Flying Training School. The air taxi side of the business remains in the ownership of Medminster and still trades as Air Gregory although it is only a shadow of what it was ten years before.

Fleet:

Piper PA24 Comanche 250	G-ARIE	c/n 24-1888	dd. 2.63,	sold 11.63.	
Piper PA27 Aztec 250A	G-ASHH	c/n 27-63	dd. 3.63,	sold 12.67.	
Piper PA27 Aztec 250B	G-ASRI	c/n 27-2352	dd. 5.64,	sold 5.71.	
	G-ASFG	c/n 27-2311	dd.10.72.		
Piper PA30 Twin Comanche	G-ASMR	c/n 30-227	dd. 2.64,	sold 8.71.	
	G-ASON	c/n 30-312	dd. 4.64,	sold 9.71.	
	G-ASYK	c/n 30-573	dd.11.64,	sold 5.70.	
	G-ASYO	c/n 30-570	dd.11.64,	sold 9.70.	
	G-AXAV	c/n 30-1793	dd.11.71,	sold 11.74.	
Piper PA27 Aztec 250E	G-AZYU	c/n 27-4601	dd.11.74		
Piper PA39 Twin Comanche	G-AYKO	c/n 39-80	dd.10.70,	sold 5.71.	
Pilatus PC6 Porter	HB-FAG	c/n 527	dd. 9.63,	sold 4.64.	
Aero Commander 500	G-ASIO	c/n 1113-55	dd.12.64,	sold 4.68.	
DH104 Dove 1	G-ALBM	c/n 04170	dd.10.68,	sold 1.70.	
	G-ANMJ	c/n 04006	dd. .68,	sold 12.69.	
	G-AOVY	c/n 04009	dd.10.68,	wfu. 1.69.	
Douglas C47A Dakota	G-AKJH	c/n 13164	dd. 4.65,	sold 10.68.	
HS.125 Series 1	G-ASNU	c/n 25005	dd.12.65,	sold 4.69.	
HS.125 Series 3B	G-AVOI	c/n 25128	dd. 5.68,	sold 9.69.	
HS.125 Series 1B	G-AWUF	c/n 25106	dd.11.68,	sold 1.69,	'Pegasus 1'.
Cessna 337A Super Skymaster	G-ATPU	c/n 0397	dd. 8.70,	sold .72.	
Cessna 310Q	G-AYND	c/n 0110	dd.11.70,	sold 5.71.	
Cessna 401	G-AWXM	c/n 0165	rd.10.70,	sold 6.73.	

AIR INTERNATIONAL

Air International Charter Ltd. Base: Stansted Airport.

Formed in May 1971 by Captain J.H.Malanowicz, Air International Charter took delivery of a Viscount 700 from Field Aircraft Services in July 1971. The aircraft was then flown, via the airline's base at Stansted, to Cardiff where the airline undertook a month's crew training in the hands of Cambrian Airways, and early in September the Viscount started charter operations from Stansted. The company had been set up to operate a series of inclusive tour services on behalf of Arrow Travel, and on September 4 1971 the Viscount operated its first revenue earning service when it flew from Stansted to Frankfurt and Vienna with a party of tourists. For the remainder of 1971 and part of 1972, Air International continued to operate these services, mainly to continental destinations, although several other ad-hoc passenger and freight charters were flown by the company to many British Airports. These flights included several on behalf of oil companies - Air International were the first airline to fly a Viscount service from Norwich Airport when G-APPX flew an oil rig crew to Esbjerg in Denmark, and a charter was also flown from Gatwick to the Shetland Islands during the autumn of 1972.

For the 1972 season the airline had hoped to purchase a Boeing 707-321 for use on its charter services, and also to start affinity group charter flights across the North Atlantic. However, these plans came to nothing, and during the summer the Viscount operated most of its flights from Gatwick and Stansted airports. Most of these flights were passenger charters to countries in Central Europe including Germany and Austria, but early in November 1972 the Viscount was impounded at Gatwick by the British Airports Authority for non-payment of landing fees. The aircraft was released later in November, and it limped to Castle Donington on three engines on November 21, was returned to its owners, Field Aircraft Services, and did not fly again.

Early in 1973, Air International announced that it was obtaining two BAC 1-11 400s for use on inclusive tour services for the 1973 season, but these plans came to nothing. The airline then announced a series of flights from Southend for the 1973 summer season using its Viscount - these flights would have been operated for local travel agencies. The company intended to fly from Southend to Palma, Munich, Beauvais (all twice weekly), Dusseldorf, Ostend, Frankfurt (all thrice weekly) and Alicante once weekly. However, Air International operated no services whatever during 1973, and in May the company ceased trading, being officially wound up on June 11 1973.

Fleet:

Vickers 702 Viscount	G-APPX	c/n	073	dd. 7.71,	ISOC, 'Toniko'.

AIR KRUISE

Air Kruise Ltd.
Air Kruise (Kent) Ltd.
Trans Channel Airways Ltd.
Skyfotos Ltd.

Bases: Lympne, Ramsgate, Lydd and Blackbushe Airports.

Air Kruise was formed in the summer of 1946 by Wing Commander H.C.Kennard, and started pleasure flying and charter flying from Lympne airfield with a Miles Messenger and Percival Q 6. Wing Commander Kennard had served throughout the 1939-45 war in Fighter Command, and he now took up the position of chairman and Managing Director of Air Kruise. In February 1947 an Airspeed Consul was added, and during the summer months, in addition to pleasure flying at Lympne, Air Kruise operated many cross-Channel charter flights to towns in Northern France and Belgium. These services continued over the next three years, and in April 1950 the company acquired its first Dragon Rapide, and on July 15 1950 this aircraft opened a thrice daily service from Lympne to Le Touquet. This service proved to be extremely popular, a second Rapide was soon acquired to help boost the capacity on these cross-Channel services, and Captain P.Cowell was appointed chief pilot.

The company also took over the operation of Ramsgate Airport, and reopened this airfield on June 27 1953 under a lease from Ramsgate Corporation. From Ramsgate, Air Kruise operated scheduled passenger services through Lympne to Le Touquet and Ostend from June onwards. These services were flown with Dragon Rapides under the title of Trans Channel Airways, and the aircraft carried both Air Kruise and Trans Channel Airways' titling At Ramsgate Air Kruise set up the Ramsgate Flying Club using an Auster Autocrat and a Tiger Moth, while another associate company, Skyfotos, used an Autocrat for aerial photographic work from Ramsgate. The Messenger was still used by Air Kruise for pleasure flying, and the aircraft mainly flew from Ramsgate on this task. However, Lympne still remained as Air Kruise's main base and during 1953 the company was taken over by the British Aviation Services group who also operated services from Lympne under the guise of Silver City Airways. Air Kruise's operations, though, continued unchanged except for the transfer of two Dakotas to the company for the operation of its scheduled services. Wing Commander Kennard remained with Air Kruise, and also joined the board of directors of British Aviation Services Ltd. By 1954 Air Kruise was flying scheduled passenger services from Lympne to Ostend, Le Touquet, Basle, Birmingham, and Zurich and during the year over twenty five thousand passengers were carried on these services.

The acquisition of further Dakotas in 1955 led Air Kruise to expand its inclusive tour charter services which were now flown on behalf of several leading tour operators. By 1955 all former Lympne services were being flown from Lydd and, in addition to the scheduled services to Le Touquet and Ostend, Air Kruise was flying summer services to Basle, Geneva, Turin, Venice, Salzburg, Copenhagen and Lyon. Inclusive tour charter flights were also made from Manchester to Ostend, and from Birmingham to Ostend and Basle using Dakota aircraft. The Dakotas were also used for a limited amount of passenger and freight charter work, and this included the transport of many Hungarian refugees to this country during 1955, after the Hungarian revolution. Air Kruise's 1956 operations were further expanded by the transfer of three Bristol Wayfarers from Silver City, and these aircraft operated alongside the Dakotas on many of Air Kruise's inclusive tour services to the Continent which by now were the most extensive of any British airline. One of the largest programmes flown by Air Kruise was on behalf of Blue Cars (Continental) Coach Cruises - using Dakotas these services were flown from Lydd to Madrid, Barcelona, Palma, Turin, Venice, Basle, Salzburg and Jersey. In addition, numerous other flights were operated to other destinations including Perpignan, Copenhagen and Geneva on behalf of other tour companies. Services were also flown from Blackbushe, and Air Kruise's operations continued throughout 1957 and into 1958, but during that year the aircraft were gradually transferred to Silver City Airways, and the name of Air Kruise disappeared.

Fleet:

Miles M.38 Messenger 2A	G-AHZS	c/n	6331	dd. 8.46,	ISOT.
Percival Proctor 1	G-AHVK	c/n	H.58	dd. 9.46,	sold 6.51.
Percival Q.6	G-AFIX	c/n	Q.31	dd. 9.46,	sold 1.49.
Airspeed AS.65 Consul	G-AIUS	c/n	750	dd. 2.47,	sold 3.54.
Miles M.65 Gemini 1A	G-AKDJ	c/n	6448	dd. 8.47,	sold 7.51.
	G-AJWH	c/n	6293	dd. 2.51,	sold .57.
Auster J/1 Autocrat	G-AIZZ	c/n	2234	dd. 9.49,	ISOT.
DH89A Dragon Rapide	G-AESR	c/n	6363	dd. 3.53,	written off Gerdes El Adol 22.9.56.
	G-AEWL	c/n	6397	dd. 4.50,	sold 1.56, 'Nicole'.
	G-ALWK	c/n	6856	dd. .50,	sold 4.57.
Douglas C47B Dakota	G-AMYV	c/n	32943	dd. .53,	ISOT.
	G-AMYX	c/n	33042	dd. 2.56,	ISOT.
	G-AMZB	c/n	26980	dd. .55,	ISOT

Douglas C47A Dakota	G-ANLF	c/n 11979	dd. 4.55,	ISOT.
Douglas C53D Dakota	G-AOBN	c/n 11711	dd. 4.55,	ISOT.
Bristol 170 Freighter Mk.21	G-AIFM	c/n 12773	dd. 2.56,	ISOT, converted to Wayfarer .56.
	G-AIME	c/n 12795	dd. 3.56,	ISOT, converted to Wayfarer .56.
	G-AHJI	c/n 12741	dd. 1.56,	ISOT, converted to Wayfarer 3.56.

AIRLINKS

Airlinks Ltd.
Machinery Finance (Overseas) Ltd.

Base: Croydon Airport.

In mid-March 1946, the Bata Shoe Company bought an Airspeed Consul to fly its executives around the country to its various factories, and also to the Continent on business. However, on April 29 1947 an airline named Airlinks was formed by Messrs.G.Knott, Wilfred E.Ridsdale and Edward G.S.Bach. This new company took over the operation of the Consul, and in addition to flying Bata's executives around the British Isles and Europe, Airlinks also offered the aircraft for executive charter and air taxi work from May 1947 onwards. In July a de Havilland Dove was added, and this increased the airline's amount of charter work. The Consul was sold in December 1948, but charter operations continued with the Dove unchanged, and on October 6 1950 Airlinks changed its name to Machinery Finance (Overseas) Ltd., and this resulted in the sale of the Dove in January 1951, and no further flying took place.

Fleet:

Airspeed AS.65 Consul	G-AGVY	c/n 3204	dd. 3.46,	sold 12.48.
DH104 Dove Series 2	G-AJGT	c/n 04034	dd. 7.47,	sold 1.51, 'Bata'.

AIR LONDON

Air London Ltd.
Airways Training Ltd.

Base: Gatwick Airport.

Airways Training was formed in 1965 by Mr. and Mrs.A.P.Mack. Based at Gatwick, this company operated Twin Comanches and a Piper Apache on pilot training to instrument rating and twin rating standards. Late in 1966 a subsidiary was formed named Air London and, using additional Beagle 206s, this company rapidly expanded - flying executive charters, air ambulance services and light freight charters. Air London was the first British air taxi operator to use the Beagle 206, and during its career with Air London this fleet of aircraft operated many unusual charters. These included a series of flights, carrying up to two thousand day old chicks, from Gatwick Airport to towns in Northern France, including St.Brieuc and Quimper, on behalf of British United Turkeys. These flights were operated for several years, and one aircraft (G-ATKP) was specially allocated for these services, flying as the 'Steam Chicken Line'.

On January 1 1969 the company bought out a two thirds share in the Shoreham Aviation group of companies. These companies included the Shoreham School of Flying, Shoreham Aviation and Shoreham Aviation Services. The other one third share of these companies was retained by Mr.Frank Hewitt of Hewitt's Investments. However, the operations of Air London continued unchanged and during January 1970 a comprehensive air ambulance service was set up. Using the Beagle 206s,there was accommodation for a nurse and two other passengers, and Air London had an agreement with a nursing organisation enabling the company to offer a complete escort service - the air ambulance, nurses, doctors and road ambulances throughout Europe. This service was available on a twenty four hour basis, and Air London's Air Operator's Certificate covered the company for these operations throughout Europe and North Africa. However, in September 1970 the fleet of Beagle 206s was retired, and a fleet of Piper Aztecs and Twin Comanches was bought as a replacement. These aircraft have since then concentrated on air taxi and training flights from Gatwick, and the Twin Comanches carry out many instrument training flights from Stansted and Gatwick.

The aircraft now fly regularly to many destinations on the Continent including Paris, Le Havre, Brussels, Antwerp, Deauville, Jersey, Lille and Cherbourg. In November 1973 Air London started operating regular air taxi flights from Rochester on behalf of Elliot Automation. The number of these flights increased from February 19 1975 onwards - this was after one of Elliot's own aircraft force landed on the M2 Motorway shortly after take-off from Rochester. Since February, therefore, Air London's aircraft have been used regularly for flights from Rochester to Warton and Woodford. Air London also operates a weekly flight, on Thursdays, from Farnborough to Warton and return. Air Ambulance flights are still operated, and in November 1974 the airline agreed a joint plan with Baylee Air Charter for the operation of emergency medical flights between the Channel Islands and the mainland. Air London's one entry into the executive jet charter market was a short one, and after only five months of operations from Gatwick the aircraft was sold to McAlpine Aviation. During its brief service with Air London the HS125 was little used, although it did make several international flights to various European cities. However, today Air London's two major activities are firstly the operation of air taxi flights from its base at Gatwick Airport, and secondly the training of pilots for Instrument Ratings at Gatwick and Stansted.

Fleet:

Beagle 206 Series 1	G-ATDD	c/n 013	dd.12.66,	sold 2.68.
	G-ATHO	c/n 019	dd.10.70,	sold 4.73.
	G-ATKP	c/n 026	dd.12.68,	sold 9.70, 'Stardust'.
	G-ATYC	c/n 039	dd. 3.67,	sold 9.70, 'Starlight'.
	G-ATZO	c/n 044	dd.12.67,	sold 9.70.
Piper PA30 Twin Comanche	G-ASRU	c/n 30-376	dd. 6.71,	sold 4.72.
	G-ATEN	c/n 30-544	dd. 5.72.	
	G-ATSE	c/n 30-983	dd. 7.74.	
	G-ATZU	c/n 30-1271	dd.11.71,	sold 7.74.
Piper PA27 Aztec 250B	G-ASRE	c/n 27-2520	dd. 4.73.	
Piper PA27 Aztec 250C	G-ATFF	c/n 27-2898	dd.10.73.	
Piper PA27 Aztec 250D	G-AXZP	c/n 27-4464	dd. 4.71,	sold 5.75.
Piper PA27 Aztec 250E	G-AZIF	c/n 27-4633	dd.11.71,	written off Stansted 5.1.72.
HS125 Series 400	G-AYNR	c/n 25235	dd. 1.74,	sold 6.74.

AIRMOTIVE

Airmotive Company Ltd.
Airmotive (Liverpool) Ltd.

Bases: Pwllheli Aerodrome, and Liverpool (Speke) Airport.

Airmotive (Liverpool) was formed at Liverpool Airport in January 1950 by Captain James Henry Hoggart-Hill and air charter operations began during the same month from Liverpool with one Dragon Rapide aircraft. Captain Hoggart-Hill became Chairman and Managing director of this company - he had originally learnt to fly in 1929 and during the war had served as Chief Production test pilot of the Westland Aircraft Company. After the war he had flown for Air Contractors, and during 1949 he was airport manager and Chief Pilot of Butlin Aviation, but at the end of the 1949 pleasure flying season he left Butlins to start his own company.

With one Dragon Rapide Air Motive was mainly concerned with pleasure flying, and the operation of army co-operation flights from Liverpool Airport. A sales office was also opened in London, and from its two bases the company also acted as aviation consultants, and as suppliers of aircraft and aero-engines, accessories and equipment. The Rapide was sold in June 1952 and all air charter operations were suspended followed shortly afterwards by the winding up of Airmotive (Liverpool) Ltd. However, the Airmotive Company continued to operate from its London offices for many years afterwards acting as aviation consultants, and also dealing in the sale of aircraft and in their ferrying and flight testing.

Fleet:

DH89A Dragon Rapide	G-AFOI	c/n	6450	dd. 1.50, sold 6.52.

AIR NAVIGATION AND TRADING

Air Navigation and Trading Ltd.
Thorne Aviation Ltd.

Bases: Carlisle (Kingstown) and Blackpool Airports.

Air Navigation and Trading was formed by Mr.Russell L.Whyham at Blackpool Airport in March 1947. Mr.Whyham was formerly a test pilot with the Royal Air Force during the war, and before the war he had been Chief Engineer with Allied Airways. Passenger and freight charters were initially operated with a Dragon Rapide, but in February 1949, several ex Royal Air Force Avro Ansons were purchased for conversion to civilian standards. However, only one of these aircraft was converted, and this was sold immediately to Starways. During 1948, Air Navigation and Trading was granted a B.E.A. Associate agreement for the operation of a scheduled passenger service between Carlisle and the Isle of Man, and this service was flown in conjunction with two other airlines. The fleet of Rapides gradually built up, and numerous charter flights were operated to the Isle of Man from Blackpool and Carlisle. At Carlisle's Kingstown Aerodrome, Air Navigation had opened a base, and several of the airline's aircraft were then based there.

In 1949, an associate company named Thorne Aviation had been started. This company was run by Wing Commander J.R.Grice, D.F.C., and all operations were carried out jointly with Air Navigation and Trading. Thorne Aviation's first work was the operation of a Fairchild Argus on pleasure flights from Blackpool during the summer months, and during 1950 this aircraft was joined by an Auster, and both aircraft were almost solely used for joy riding. However, in 1951 Air Navigation purchased the two surviving de Havilland Dragons and after overhaul, these two aircraft replaced Thorne Aviation's Auster and Argus on pleasure flying. These Dragons operated flights around Blackpool Tower for ten years in the livery of Air Navigation and Trading before being retired in 1962. Three Avro 19s were purchased in 1950, and these were used on passenger charters to the Isle of Man, and at least one flew in the livery of Thorne Aviation. During 1951 the airline operated scheduled services under a B.E.A. Associate agreement, from both Carlisle and Blackpool to the Isle of Man with the Rapides. In January 1959 the airline applied to the A.T.L.B. for permission to operate a Blackpool to London scheduled service and at this time instructed de Havilland (Australia) to modify a Drover for use on this service. However, the service was not approved.

Air Navigation also operated a first class maintenance service at both Carlisle and Blackpool Airports. C of A overhauls, and the servicing of aircraft, engines and ancillary equipment was undertaken, and contract work was also carried out for the Government, and air ambulance work was also undertaken. However, the base at Carlisle was later closed, and all work was transferred to Blackpool. When the Dragons were retired in 1962, pleasure flying continued using Dragon Rapides, and for the 1962 season only, a Percival Prentice. During the late 1950s, the company had acquired several unusual aircraft for conversion, for use on passenger services, but of all the Marathons, Drovers and a single Pembroke acquired, none ever flew for the company. During the 1960s, Air Navigation built up a large fleet of training aircraft including many Piper and Cessna single-engined aircraft, and by 1969 the only charter work carried out by the company was flown using a Cherokee Six, and this was mainly used for pleasure flying. However, in July 1974, a Twin Comanche was bought for air taxi work out of Blackpool, and this aircraft is a regular visitor to Ronaldsway, and several trips have also been made to London, and some to Scotland.

Fleet:

Fairchild Argus 2	G-AJBF	c/n	869	dd. .49, sold 10.51, operated by Thorne Aviation.
	G-AKJA	c/n	302	dd. 7.49, sold 7.52.
	G-AKCJ	c/n	306	ca. 7.49, sold 9.51.
DH89A Dragon Rapide	G-ACPP	c/n	6254	dd. .60, sold 6.61.
DH89A Dragon Rapide	G-AHJA	c/n	6486	bt. 8.63, sold 6.67.
	G-AJBJ	c/n	6765	bt. 7.55, wfu. 9.61.
	G-AJKW	c/n	6539	dd. 4.63, sold 3.65.
	G-AKOY	c/n	6504	dd. .48, sold .50.
	G-AKSG	c/n	6931	bt. 8.49, sold 4.54.
	G-AKZT	c/n	6894	bt. 5.48, sold 4.52.
	G-ALXI	c/n	6690	bt. 5.50, sold 7.55.
	G-ALXJ	c/n	6863	dd. 4.50, written off Isle of Man 10.7.51.
	G-APSD	c/n	6556	dd. 2.59, sold 7.63.
Auster 4	G-ALVV	c/n	865	dd.10.50, sold 6.53.
Auster J/2	G-AJPU	c/n	2392	rd. 9.55, sold 8.56.
Avro 19 Series 1	G-AHIB	c/n	1317	bt. 4.50, sold 6.57.
	G-AHIE	c/n	1320	bt. 4.50, wfu. 3.59.
	G-AHIG	c/n	1322	bt. 8.50, sold 4.55.
Avro 652A Anson 1	G-AMDA	ex.	N4877	dd.10.53, sold 7.55.
Avro 652A Anson 11	G-ALIH	ex.	NL229	dd. 2.49, sold 6.51.
DH84 Dragon 1	G-ACIT	c/n	6039	dd.11.51, sold 1.62.
DH84 Dragon 2	G-ADDI	c/n	6096	bt. 2.51, sold 11.62.
DH86B	G-ADUF	c/n	2334	rd. 1.51, sold 10.51.
Percival Proctor 4	G-AJMP	c/n	H.742	dd. .50, wfu. .53.
DHA.3 Drover	G-APPP	c/n	5002	never arrived from Australia.
	G-APXX	c/n	5014	dd. 9.61, never entered service.
Percival Pembroke	G-APNL	c/n	13	dd. 6.58, never entered service.
Miles M.60 Marathon 1A	G-AMHS	c/n	130	rd. 9.58, never entered service.
	G-AMHV	c/n	133	dd. 6.60, sold 3.61.
Percival Prentice	G-APJE	ex.	VS282	dd. 3.62, sold 9.62.
Aermacchi AL60B-1	G-ARZG	c/n	6219-39	dd. 8.62, written off Squires Gate 7.7.65.
Beagle A61 Terrier 2	G-ASAN	c/n	B.608	dd. 6.63, sold 5.64.
Aero Commander 520	G-ARJJ	c/n	16	dd. 5.64, sold 10.66.
Piper PA30 Twin Comanche	G-ATZU	c/n	30-1271	dd. 7.74.

AIR PEGASUS

Air Pegasus Ltd.

Bases: Initially Cruchfield Manor, Maidenhead, later Preston Candover, Hampshire.

Air Pegasus was formed late in 1972 by Mr.Peter Cadbury as an air taxi and executive charter operator. The fleet was based at Mr.Cadbury's house at Cruchfield Manor near Maidenhead, and initially consisted of an Islander, two Senecas, an Aztec and a Jetranger. Air taxi operations continued from Cruchfield Manor for over eighteen months, and many of Air Pegasus' services were flown from Heathrow and Gatwick Airports. However, early in 1974 Mr.Cadbury moved house to Preston Candover in Hampshire, and a new airstrip was built here, and this is now Air Pegasus' headquarters. Many of the company's services are now flown out of Heathrow, Northolt and Blackbushe - many of the flights are international charters to the Continent, and Air Pegasus also flies many services to airfields in Scotland. A Cessna Golden Eagle was acquired in November 1974 for contract work, and although the original work for the aircraft fell through, the aircraft was retained and during 1975 this aircraft flew many executive charters to destinations abroad including Paris, Amsterdam and Dusseldorf. During 1975 one of the Aztecs (G-BBHF) was based at Rotterdam and operated under contract to the Dutch firm of Typssoos and Partners. In addition many flights were operated to racemeetings, both in this country and abroad, flying jockeys, owners and trainers to the meetings. Air Pegasus' aircraft are, therefore, familiar sights at many aerodromes both at home and abroad particularly at many of the London area airfields and also at Birmingham, Le Touquet, Norwich, Ostend and Teesside.

Fleet:

BN.2A Islander	G-AZEI	c/n 291	dd.11.71, sold 1.74.
Piper PA34 Seneca 200	G-AZOT	c/n 34-50073	dd. 2.73, sold .74.
	G-BAIG	c/n 34-50243	dd.10.72, sold 4.74.
	G-BAMZ	c/n 34-50035	dd.11.72
Piper PA28 Cherokee 180	G-BASJ	c/n 28-5136	dd. 4.73, sold 9.73.
Piper PA27 Aztec 250E	G-BATE	c/n 27-5006	dd. 3.73, sold 11.73.
	G-BAWY	c/n 27-5049	dd. 6.73, sold 12.74.
	G-BBHF	c/n 27-5166	dd.11.73.
	G-BBRJ	c/n 27-5223	dd. 7.74.
Augusta Bell 206B Jetranger 2	G-BAKU	c/n 8342	dd. 3.73.
Cessna 340	G-BAYK	c/n 0156	dd. 4.74, sold 9.74.
Cessna 421A	G-AWBK	c/n 0125	dd. 8.74, sold 11.74.
Cessna 421B	G-BBSU	c/n 0605	dd.11.74.

AIR SAFARIS

Air Safaris Ltd.
African Air Safaris Ltd.
Meredith Air Transport Ltd.
Don Everall Aviation Ltd.

Bases: Southend later Gatwick later Bournemouth and Birmingham Airports.

Air Safaris' history can be traced back to September 21 1950 when a company named Tropic Airways was registered in South Africa as an airline company. This charter firm was formed by Captain Tom Meredith, Captain H.Creed and Captain W.E.Hamilton, D.F.C. D.F.M. to operate charter flights between South Africa and Europe. On March 18 1951 Tropic's Dakota completed its first flight to London from South Africa - the aircraft had taken a ship's crew to Hamburg, and then positioned to London before flying a group of German passengers to Johannesburg. During the next twelve months, this charter company operated flights to South Africa from Hamburg and Amsterdam carrying German and Dutch immigrants. A certain amount of ad-hoc charter flying was also undertaken including the transporting of ship crews and machinery.

In 1952 the company purchased an ex Royal Air Force Dakota, and in order to enable Tropic to conform with the conditions of sale imposed on the aircraft by the Air Ministry, a British company had to be formed. This company was registered on August 12 1952 as Meredith Air Transport, and one of the company's first charters was to fly a consignment of radio parts to Khartoum for International Aeradio in December 1952. At the end of December the aircraft was used on a fourteen day Eastern Mediterranean inclusive tour, and during 1953 the aircraft was used for a series of air cruises around the Mediterranean organised by E.A.Gibson. Also during 1953, the charters to South Africa continued from London, but in June the Dakota was sold in order to raise further capital, and to pay off outstanding debts. During 1953 Tropic's founders left the company and formed Trek Airways, and this airline operated similar services to those of Tropic from August 1954 onwards with a Vickers Viking acquired by Meredith Air Transport. Trek bought two more Vikings from Airwork in October 1954 and November 1955, and these aircraft were flown under South African registry. Meredith Air Transport, meanwhile, continued in existence but had no aircraft, and on November 29 1954 it changed its name to African Air Safaris. In October 1956 African Air Safaris bought a B.O.A.C. Hermes for Trek's services between South Africa and Europe, but the aircraft was not suitable for these services and was immediately leased to Britavia. However, when Trek introduced DC4s onto its services in 1958, two Vikings were transferred to African Air Safaris, and these aircraft flew low cost tourist services from Southend to Johannesburg during 1958. African Air Safaris remained as Trek's British sales agent, and during 1958 and 1959 the airline started to operate inclusive tour flights and other charters to Scandinavia and to many Mediterranean resorts. During the 1959 summer season African Air Safaris flew inclusive tour services from Gatwick to Basle, Bergen, Dusseldorf, Geneva, Jersey, Lyon, Munich, Nice, Palma, Perpignan, Rimini, Rotterdam, Tarbes, Toulouse and Zagreb with the two Vikings in addition to numerous services from Southend to the Continent. Many ad-hoc charters were also flown - on March 13 G-AHOW flew a passenger charter from Dusseldorf to Stansted, while the return trip was made by G-AHOR on March 16. The 1959 season was, therefore, a busy one and on November 26 1959 the airline changed its name to Air Safaris, and with an increased amount of inclusive tour charters planned from Gatwick for the 1960 season, Air Safaris moved its operations base to Gatwick Airport late in 1959.

By 1960 the services to South Africa had come to an end, and early in the year the company set up its maintenance base at Hurn Airport, Bournemouth. 1959's fleet of two Vikings was increased to four Vikings and one Hermes, and with these aircraft Air Safaris flew a large number of inclusive tour services during the 1960 summer season. From its maintenance base at Hurn, the airline flew many charters to Amsterdam, Jersey and Le Bourget with the Vikings, while the Hermes flew occasional services to Tarbes, Palma and Treviso, but most of its operations were conducted from Gatwick Airport. From Gatwick the company operated a large inclusive tour programme with both the Vikings and the Hermes, and there were services to Barcelona, Basle, Copenhagen, Dusseldorf, Frankfurt, Hamburg, Innsbruck, Le Bourget, Luxembourg, Lyon, Munich, Nice, Palma, Perpignan, Pisa, Rome, Rotterdam, Tarbes, Toulouse and Zurich. Occasionally Air Safaris chartered Falcon Airways Hermes aircraft for some of its services - on August 15 1960 Air Safaris' chief pilot, Captain C.M. de Bounevialle, flew Falcon's Hermes G-ALDA from Hurn to Spain and return on an inclusive tour charter flight. Ad-hoc charter work was still undertaken by the company, and one of the more interesting flights of the year was made by Viking G-AJFT on July 28 1960 when it flew a charter from Gatwick to Budapest. In November 1960 the company bought out the air transport division of Don Everall Aviation, and also the remains of the Falcon Airways Hermes fleet. From Falcon, Air Safaris bought one flyable Hermes and another unserviceable aircraft for spares use - the flyable aircraft (G-ALDA) was delivered to Gatwick on November 23 and operated its first service for Air Safaris later that day when it flew a passenger charter to Frankfurt. From Don Everall, Air Safaris inherited a network

G-AKJH Air Gregory Douglas DC-3 — Ian MacFarlane

G-ASON Air Gregory Twin Comanche — Austin J Brown

G-AWUF Air Gregory HS.125 — Transportraits/W Lewis

G-ANMJ Air Gregory Dove — Ian MacFarlane

G-ASIO Air Gregory Aero Commander — Ian MacFarlane

G-ARZG Air Navigation & Trading AL.60 — Ian MacFarlane

One of the shortest lived of all Viscount operators was Air Ferry, who only used these aircraft for one season. G-AVHE pictured here at Newcastle on July 11, 1968.

Ian MacFarlane

G-APPX was Air International's only aircraft, but its orange and white livery was seen at many airfields during 1971 and 1972.

Austin J Brown

Air Kruise used three Bristcl Wayfarer, of which G-AIME is depicted here, for its tour flights to the Continent from Lydd and Blackbushe.

Arthur Pearcy Jnr.

G-AESR Air Kruise Dragon Rapide A J Jackson

G-AOBN Air Kruise Douglas DC-3 Ian MacFarlane

G-ATKP Air London Beagle 206 S G Jones

G-ASRE Air London Aztec Tony Merton Jones

G-ATEN Air London Twin Comanche Tony Merton Jones

G-BBHF Air Pegasus Aztec Tony Merton Jones

of both scheduled and inclusive tour services. The schedules served Exeter, Guernsey, the Isle of Wight and Jersey from Birmingham, while the inclusive tours served Basle, Malaga, Nice, Perpignan and Rimini from Birmingham. The company also inherited a Birmingham to Newcastle licence, a route which Don Everall had never flown, and it was this service that Air Safaris inaugurated on June 3 1961 with Viking G-AHOW flying from New-Newcastle, via Birmingham, to Bournemouth. Prior to this, in April 1961, Air Safaris had opened its first scheduled service - a twice weekly Bournemouth to Amsterdam via Ostend service flown by Vikings. A Jersey to Ostend and Rotterdam service, applied for in September 1960, was never flown by the airline.

1961 was a year of major and rapid expansion for Air Safaris, and early in the year the company received a demonstration of the Dart Herald, and in subsequent licence applications for inclusive tour services, the airline specified the use of both the Herald and the Avro 748. Following the introduction of the service to Ostend and Amsterdam, the airline introduced a Viking service from Bournemouth to Dublin and Belfast later in April. On May 27 the Birmingham to Jersey service was reopened, with a Coventry to Jersey service starting on June 30, and a weekly service from Birmingham to Guernsey being opened on July 1. During the summer, the Newcastle to Bournemouth service was extended to Exeter, and on June 3 a Birmingham to Sandown service was started using de Havilland Herons on lease from North South Airlines. While the airline was using its Vikings to expand its scheduled services, a fleet of Hermes was being used to build up a large network of inclusive tour services. These aircraft, and some of the Vikings, were flying holidaymakers from Birmingham, Bournemouth, Gatwick, Glasgow, Manchester and Newcastle to numerous Mediterranean sunspots. From Newcastle, Air Safaris flew services to Barcelona, Palma, Perpignan, Rimini, Tarbes and Valencia on behalf of Horizon Travel and Airway Holidays. The first service from Newcastle was flown on May 21 by Hermes G-ALDL with a flight to Palma, but when loads were light Vikings were used, with G-AHPC operating a Newcastle-Manchester-Valencia flight on June 5. From Manchester, Air Safaris flew services to Barcelona, Bordeaux, Caen, Copenhagen, Frankfurt, Palma, Rimini. St.Nazaire, Tarbes and Valencia. During the winter of 1960 Air Safaris continued to operate many of Don Everall's almost daily charters from Birmingham to Le Bourget and Dublin, but when the summer of 1961 came, Air Safaris concentrated on the operation of its inclusive tours and scheduled services from Birmingham - the inclusive tours served Copenhagen, Perpignan and Stavanger.

The services from Renfrew Airport, Glasgow, were mostly flown via Manchester, Hurn or Gatwick as the loads generally did not justify the operations of direct services from Glasgow, and the flights were operated via other airports in order to collect further passengers. However, Air Safaris did operate a direct Ostend to Renfrew service on Sundays fortnightly. From Hurn the airline flew regularly to Copenhagen, Le Bourget, Palma, Perpignan, Tarbes and Valencia. Air Safaris' largest inclusive tour charter programme in 1961, though, was flown from Gatwick and destinations served included Barcelona, Basle, Copenhagen, Dinard, Dusseldorf, Frankfurt, Innsbruck, Le Bourget, Lisbon, Luxembourg, Lyon, Maastricht, Malaga, Milan, Munich, Nice, Palma, Perpignan, Rome, Santander, Tarbes, Toulouse and Valencia. These services were mainly flown on behalf of Horizon Travel and Universal Sky Tours, and between July and September 1961 Air Safaris applied for one hundred and twenty 'B' licences for the operation of inclusive tour services during the following winter. Before the Air Transport Licensing Board would authorise the operation of these services, however, they required that Air Safaris' financial resources be strengthened, and at that time credit was difficult to obtain following the collapse of Overseas Aviation in August. Air Safaris had flown many of Overseas' tours during August and September, but Air Safaris, too, was suffering from financial problems, and gradually the services were cut back. The last Bournemouth-Birmingham-Newcastle service was flown by Viking G-AGRS on September 4, while the last inclusive tour flight from Newcastle was operated by Hermes G-ALDE on October 8. During October the services were gradually pruned, although some of the remaining inclusive tour charter flights were operated by leased Tradair Vikings and Viscounts. The last scheduled Hurn to Dublin service was flown by Viking G-AHOW on October 27, and during the next two days only a handful of charter flights were operated to Barcelona, Dusseldorf and Le Bourget. On October 30 Hermes G-ALDA flew a service to Tenerife from Gatwick, while Viking G-AJBX flew from Gatwick to Gibraltar - G-AJBX returned the following day to Gatwick, and later departed for Copenhagen, but at 19.11 on October 31 Hermes G-ALDA landed at Gatwick after a flight from Lisbon, marking the operation of Air Safaris' final service.

The airline was, sadly, unable to raise any further finance, and no further flying took place - most aircraft returned to Hurn where they were either sold or scrapped. Air Safaris had debts totalling £521,073, of which over one hundred thousand pounds was owed to Esso Petroleum for fuel, and on January 11 1962 all the airline's licences were revcked.

Fleet:

Douglas C47B Dakota	G-AMSU	c/n 33548	dd.10.52, sold 6.53.
Vickers 498 Viking	G-AGRS	c/n 110	dd.10.60, ISOC.
	G-AHOR	c/n 118	dd. 4.58, written off Tarbes 29.5.60, 'Alouette 11'.
	G-AHOS	c/n 119	dd. 2.61, ISOC.
	G-AHOT	c/n 121	dd. 2.53, sold 9.54.
	G-AHOW	c/n 124	dd.10.54, sold 11.54.
	G-AHOW	c/n 124	dd. 7.58, ISOC, 'Skukuza'.
Vickers 610 Viking	G-AHPR	c/n 164	dd. 3.61, ISOC.
	G-AJBX	c/n 249	dd. 3.61, ISOC.
	G-AJCA	c/n 252	dd. 8.60, wfu. 8.61.
	G-AKBG	c/n 263	dd.11.60, ISOC.
Vickers 621 Viking	G-APAT	c/n 153	dd. 2.61, wfu. 5.61, never entered service.
Vickers 627 Viking	G-AJFT	c/n 238	dd. 5.60, ISOC.
Vickers 634 Viking	G-AKTU	c/n 211	dd. 4.61, ISOC.
Vickers 639 Viking	G-AHPC	c/n 133	dd. 6.60, ISOC.
Handley Page 81 Hermes 4	G-ALDA	c/n HP81/2	dd.11.60, ISOC.
	G-ALDE	c/n HP81/6	dd. 5.61, ISOC.
	G-ALDL	c/n HP81/11	dd. 5.61, ISOC.
	G-ALDM	c/n HP81/14	dd. 4.60, ISOC.
	G-ALDT	c/n HP81/20	dd. 6.61, ISOC.

AIR SERVICES (SUNDERLAND)

Sunderland Air Charter Ltd.
Air Services (Sunderland) Ltd.

Base: Sunderland (Usworth) Airport.

Sunderland Air Charter was formed at Sunderland Airport by Mr.S.E.Marples during the spring of 1973 with one Piper Twin Comanche aircraft. From Sunderland this Twin Comanche operated many flights to other aerodromes in Northern England, to Scotland and several flights were also made to airports on the Continent. Then, in May 1975, the company obtained a contract for the operation of regular air taxi flights from Sunderland to Amsterdam and Rotterdam. At this time Sunderland Air Charter changed its name to Air Services (Sunderland), and an Aztec was purchased to supplement the Twin Comanche on the services to the Netherlands. In addition, many other international charter flights are made from Sunderland to Scandinavia, and much of the company's work is derived from locally based shipping companies. When customs facilities are not available at Sunderland, the aircraft stage via Newcastle or Teesside on international flights. During 1975, the Aztec spent most of May and June flying services between Sunderland and Rotterdam, although on June 7 this aircraft flew a service from Brest to Rotterdam, and occasionally the Twin Comanche was used on the flights to Rotterdam. Air Services Sunderland's aircraft also visited many other airports during 1975 including Bembridge, Biggin Hill, Haverfordwest, Leeds,

Ronaldsway, Sumburgh and Sydenham. Regular visits are made to Teesside and Newcastle which are both near to the company's base at Sunderland Airport.

Fleet:

Piper PA30 Twin Comanche	G-AVPS	c/n 30-1548	dd. 5.73.
Piper PA27 Aztec 250D	G-AYVJ	c/n 27-4020	dd. 5.75.

AIRSPAN

Airspan Travel Ltd.
Flyaway Ltd.

Bases: Redhill and Croydon Airports.

This air charter company was formed in April 1947, as Flyaway, by B.R.W.Betts and O.B.Betts, and started operations from Redhill and Croydon with one Percival Proctor V during April. Flyaway was registered as an airline company on June 24 1947, and its charter operations were mainly carried out from Croydon where the company opened an office. In October the Proctor was joined by a Consul, and this aircraft was bought to enable Flyaway to undertake more flights to the Continent. This activity only lasted, though, until November 14 1947 when the Consul force landed to the north east of Chalons in France, and was written off. The Proctor continued its air taxi flights and pleasure flying activities from Croydon, and on April 21 1948 another company was registered with the same directors as those of Flyaway - this new company, named Airspan Travel, worked in close association with Flyaway, and was also based at Croydon.

Airspan Travel bought a Consul, and this was mainly used on passenger charter work from Croydon Then, during 1948, Airspan Travel applied to operate several scheduled passenger services in association with British European Airways - these routes included a service from Croydon to Dinard, which was flown during the summer of 1949. Services were also flown to other airports in northern France, but at the end of 1950 both Flyaway and Airspan Travel sold their two Consuls and one Proctor, and all operations were suspended. This closure was short lived, and in July 1951 another Proctor V was registered and used by Flyaway for air taxi flights. An office was also opened in Regent Street in London, but by the end of 1952 all operations had ceased completely. The name of Airspan Travel reappeared once again - in August 1954, a company of this name chartered a Lancashire Aircraft Corporation Avro York for five months, but for what purpose is not known.

Fleet:

Percival Proctor V	G-AIZC	c/n Ae.104	dd. 4.47,	wfu. 7.50.
	G-AGTE	c/n Ae.14	rd. 7.51,	sold 2.53.
Airspeed AS.65 Consul	G-AJGI	c/n 5125	dd.10.47,	written off Chalons, France 14.11.47.
	G-AJLP	c/n 5135	dd. .48,	sold 9.50.
	G-AIOX	c/n 2188	dd. 7.49,	sold 12.50, 'Gorgeous Gussie'.
Avro 685 York 1	G-AHFB	c/n 1305	lsd. 8.54,	ret. 1.55, leased from Lancashire Aircraft Corporation.

AIRSWIFT

Airswift Ltd.
Airswift (Wales) Ltd.

Base: Blackbushe Airport.

Airswift was formed at Blackbushe early in 1973 with one Piper Seneca, which was soon joined by a second aircraft. Air taxi operations are mainly undertaken from Blackbushe, and many international charters are undertaken, and the green, orange and white livery of Airswift has become a familiar sight at many aerodromes in Southern England, Northern France and Belgium. In 1974 a subsidiary company named Airswift (Wales) was formed, and this company operates a Seneca on air taxi work mainly from Haverfordwest, Swansea and Cardiff. This work is mainly in connection with the exploration of oil off the Welsh coast,but during the winter of 1974 the oil drilling work was suspended in the area and not resumed until the spring, thus causing the amount of work to fall off during this period. Airswift's aircraft are very regular visitors to racemeetings in the British Isles and Northern France, and Airswift's aircraft regularly carry jockeys, trainers and owners around the country on charter flights to racecourses including Haydock Park, Newmarket, Doncaster, Newbury and Shoreham (for Brighton races). From Shoreham, Airswift also flies many air taxi flights across the Channel to airfields in Northern France including Cherbourg, Deauville and Le Touquet. When customs are not available at Blackbushe, the company stages via Gatwick to clear customs on international charter flights and during 1975 these flights included charters to Deauville, Ghent, Jersey, Le Havre, Ostend and Toussus.

During 1975 Airswift obtained an Air Operator's Certificate to allow it to operate the British Car Auctions fleet of aircraft. These aircraft, also based at Blackbushe, included a Piper Aztec and a Navajo Chieftain, and in addition to operating these aircraft on behalf of British Car Auctions, Airswift also flew them on their own services. However, the Navajo Chieftain (G-BBPV) was tragically lost,and Airswift's chief pilot killed,when it crashed near Camberley on October 19 1975 after hitting the ground while flying in fog. The Aztec is still operated, and is also often operated by Blue-Air Flying Services, another Blackbushe based air taxi operator.

Fleet:

Piper PA34 Seneca 200	G-BAIG	c/n 34-50243	dd. 4.74.
	G-BANS	c/n 34-50021	dd. 2.73.
	G-BASX	c/n 34-50123	dd. 4.73.
	G-BBNI	c/n 34-50312	dd.10.74.

AIR TAXI

Air Taxi (Cumberland) Ltd.

Bases: Carlisle and Prestwick Airports.

Air Taxi (Cumberland) was formed in 1957 at Carlisle Airport to operate ad-hoc passenger and freight charter flights in the north of England. A Dragon Rapide was purchased for this work, and this aircraft flew to many airports in the north of England, Scotland and Northern Ireland on charter flights. However, the amount of work generated at Carlisle was insufficient to keep the aircraft based there, and by the spring of 1959 the aircraft had moved north to Prestwick Airport. From Prestwick the company flew its Rapide on numerous pleasure flights from the public enclosure on almost every day of the week. In the capable hands of Captain Wood, this aircraft would spend many hours of every day carrying people aloft for, perhaps, their first flight in an aeroplane.

Although this joy riding took up most of Captain Wood's time, he, and the aircraft, did manage to undertake several ad-hoc charter flights. From Prestwick the Rapide would often make the short journey to Renfrew - one such flight being on July 20 1959, before making the picturesque journey onto Benbecula from Renfrew. Most of Air Taxi's flights were over some of the most magnificent scenery that any of Britain's airlines ever flew over, and during the summer months in particular, Air Taxi would make many journeys to outlying islands off the shores of Scotland. The Rapide made many flights to Benbecula, Kirkwall and Islay and several other Scottish airfields and regular charters were flown from Renfrew and Prestwick to Belfast and the Isle of Man. On August 2 1959 the

Rapide actually ventured as far as Newcastle after operating a charter flight from Belfast, but Air Taxi rarely crossed the border into England with most of its work being confined to the shores of Scotland. During the following year, the company's activities continued unchanged with many flights being made to Belfast and Renfrew, but the 1960 summer season was to be Air Taxi's last, because on April 19 1961 the Rapide once again moved south - this time to Newcastle under the ownership of Tyne Tees Air Charter. Air Taxi (Cumberland), therefore, ceased to exist, but for many thousands of tourists to the Prestwick area, the Rapide will be remembered as their first taste of real flying.

Fleet:

DH89A Dragon Rapide	G-ALPK	c/n	6757	dd. .57, sold 4.61.

AIR TAXIS

Air Taxis Ltd.
Air Taxis (Croydon) Ltd.

Bases: Croydon Airport before and after World War 2, Barton during the war.

Air Taxis was formed late in 1934, as an air charter operator, by Mr.C.W.Allen, Mr.J.D.Smith and Mr.F. Bingham. Using one Airspeed Courier, the airline flew charter passenger and freight services from Croydon to many parts of the Continent, but when war broke out in 1939, all charter operations had to be suspended. However, the airline continued in existence by undertaking aircraft maintenance and repair work at Barton Aerodrome, near Manchester, and when the war ended this work continued at Barton, carrying out C of A overhauls and aircraft repairs.

However, a new company named Air Taxis (Croydon) was registered on December 4 1946, by Mr.B.Quinn and Mr.H. Bryan, as an air charter operator. Air taxi and charter work was carried out with a fleet of Proctors, a Dragon and a Dragon Rapide, and during the next year close association was maintained with Southampton Air Services including the pooling of aircraft and operations. This pooling of aircraft included the use by Air Taxis of a Lockheed 12, Anson and Rapide on services from Paris to Croydon during June 1947 to bring back to London stranded English tourists during a French rail strike. In addition to the aircraft already mentioned, Air Taxis also leased a Pan African Airways Dakota for these services. But early in February 1948, Air Taxis (Croydon) was forced into liquidation, and no further operations took place.

Fleet:

Airspeed AS.5 Courier	G-ACLS	c/n	13	dd.10.34, written off Grenoble 17.10.34.
	G-ACLT	c/n	14	dd. 5.39, sold 3.40.
DH90 Dragonfly	G-AESW	c/n	7544	bt.12.39, sold 5.40.
DH85 Leopard Moth	G-ADWY	c/n	7117	bt. 6.38, sold 3.40.
Percival Proctor 1	G-AGYA	c/n	K.247	dd. .46, sold 2.48.
	G-AGYB	c/n	K.265	dd. .46, sold 2.48.
DH84 Dragon 2	G-AECZ	c/n	6105	dd. 4.46, sold 2.48.
DH89A Dragon Rapide	G-AGZO	c/n	6913	dd. 2.46, ISOC.

AIR TAXIS INTERNATIONAL

Air Taxis International Ltd.
Spooner Aviation Ltd.
Graysbrook Leasing Ltd.

Bases: Fairoaks Aerodrome, later Shoreham Airport.

Air Taxis International was formed in the summer of 1971 as an associate company to Spooner Aviation. Spooner's had started in business a little time before dealing in the sale of new and second hand Piper aircraft, and very quickly this company grew to be one of the largest of its type in the country. Based at Fairoaks, Air Taxis International started air taxi operations in July 1971 with a second hand Piper Aztec. During 1972, Spooner Aviation was awarded the British dealership for the Enstrom helicopter, and in September the two companies both moved base to Shoreham Airport, where new hangarage facilities have since been built. Air taxi services have used a variety of aircraft, particularly as Spooners deal in such a large number of Piper Twins, and charters have been flown by many of these aircraft, and in addition, many aircraft have been leased out to other operators. From Shoreham, many of Air Taxi's runs are to airfields in northern France including Deauville, Cherbourg and Le Touquet, and in addition several runs have recently been made to the Channel Islands, especially to Alderney. An associate company of Spooner Aviation named Graysbook Leasing, charters many of Spooner's aircraft out to air taxi operators - these charters have included the leasing of Aztecs to Edinburgh Flying Services and Glamair.

Fleet:

Piper PA27 Aztec C	G-AYWF	c/n	27-3911	dd. 7.71, sold 7.75.
	G-BAJU	c/n	27-2546	dd.10.72, sold 2.73.
Piper PA27 Aztec D	G-BADE	c/n	27-4205	dd. 8.72, sold 5.73.
Piper PA39 Twin Comanche C/R	G-AYZP	c/n	39-91	rd.10.72, sold 1.74.

AIR TRANSPORT ASSOCIATION

Air Transport Association Ltd.

Bases: Guernsey and Nottingham Airports.

Air Transport Association was formed at Guernsey Airport in November 1946, to undertake passenger and freight charter work. During the same month the company took delivery of a Percival Proctor which was used on air taxi services, and on many inter - Channel Islands flights. In January 1947 three Ansons were acquired, and the first of these aircraft entered service during that month operating many passenger charter flights from Guernsey to the South coast of England, and to Northern France. The second Anson entered service during March, and during the spring these aircraft assisted in the airlift of fresh fruit, flowers and vegetables from Guernsey to towns in Southern England. During the summer months the Ansons plied regularly between many English airports and Guernsey and Jersey carrying holidaymakers, but it was while one of the Ansons was positioning from Jersey to Southampton to collect more tourists that it crashed on July 3 1947 while letting down for an approach to Southampton. However, these services continued throughout the summer of 1947 with most of the company's flying being undertaken by the Ansons. Air Transport Association set up a subsidiary base at Tollerton Airport, Nottingham during the summer, and, in conjunction with Blue-Line Airways, Air Transport Association flew regular passenger services with the Ansons between Nottingham and the Channel Islands.

Two Republic Seabees were bought in April 1947, but they flew little, although both aircraft did aspire to gain Certificates of Airworthiness. These aircraft were sold in April 1948, and shortly after this Air Transport Association ceased operations in Guernsey and its services from Nottingham were taken over by Blue-Line Airways.

Fleet:

Percival Proctor 1	G-AILN	c/n	K.264	dd.11.46, ISOC.
Avro 652A Anson 1	G-AHFV	ex.	AW996	dd. 1.47, written off Isle of Wight 3.7.47.

Avro 652A Anson 1	G-AIWH	ex. NK787	dd. 4.47,	sold 3.48.
	G-AJBA	ex. AX409	dd. 3.47,	ISOC.
Republic Seebee	G-AJVO	c/n 645	dd. 4.47,	sold 4.48.
	G-AJVP	c/n 644	dd. 4.47,	sold 4.48.
Miles M.65 Gemini 1A	G-AKDH	c/n 6449	dd. 8.47,	ISOC.

AIR TRANSPORT CHARTER

Air Transport Charter (Channel Islands) Ltd. Bases: Jersey and Guernsey Airports, later Blackbushe Airport.

Air Transport Charter (Channel Islands) was formed at Jersey Airport in July 1946 as an air charter operator. Operations did not commence, though, until March 1947 when two Miles Aerovans were used to carry Channel Islands' vegetable and flower produce to the mainland for sale in markets in the British Isles. On April 2 1947 the airline took delivery of its first Dakota for use on passenger and freight charter work from the Channel Islands to the mainland, and to towns in northern France. This Dakota flew its first service for Air Transport Charter on April 2, and during the summer this aircraft was used to carry many holidaymakers to the Channel Islands, from airports in the British Isles as far north as Newcastle. By June the two Aerovans had been returned to their manufacturer, and replaced by three Dragon Rapides which were used on many inter-island charter flights. Close association was maintained with Island Air Charters who also operated Rapides on inter-island flights. Captain W.Gratrix of Island Air Charters later joined the Board of directors of Air Transport Charter.

By October 1947 three Dakotas were in service with Air Transport Charter, but during the winter the company was not able to find sufficient work in the Channel Islands for all three aircraft. Consequently the Dakotas were found work early in 1948 transporting fruit from French and Italian airports to Bovingdon Airport. It was while completing one of these flights that the airline lost its first aircraft in a fatal accident. Dakota G-AJBG was inbound to Bovingdon from Valencia with a cargo of fruit, on May 20 1948, when it crashed while approaching its destination in conditions of poor visibility and low cloud, killing three of the four crew members on board. This aircraft was quickly replaced by another Dakota in June, and the fleet of Dakotas was used, during the summer, for carrying many tourists to the Channel Islands for their summer holidays. However, this activity was disturbed on August 4 1948 when one of the aircraft (G-AJVZ) started flying sorties from Fassberg on the Berlin Airlift. This aircraft, in freighter configuration, operated two hundred and five sorties before being withdrawn from the airlift on November 10 1948, after clocking up a total of five hundred and sixty two flying hours. The aircraft returned to its base in the Channel Islands on November 15 to rejoin the other members of Air Transport Charter's fleet now resting after a busy summer.

The winter of 1948 was a quite time for Air Transport Charter, but once again the spring brought the usual flurry of flowers, vegetable and fruit charters to the mainland. During the summer of 1949 the Dakotas plied regularly from airports in the British Isles to the Channel Islands carrying holidaymakers, and numerous flights were operated between Blackbushe and Jersey. 1950 was a year of change for the airline - in April the company carried a party of missionaries from Amsterdam to East and Central Africa, and during the year many long distance charters were undertaken with the Dakotas. The vegetable produce flights were still flown, and apart from the normal run of trips between the Channel Islands and the mainland,an occasional flight was made from Rennes to Southampton carrying up to three and a half tons of lettuce and carrots on each trip. The Dakotas also found work flying ships' crews around Europe on behalf of ship owners - early in March, twenty seven seamen were flown from Luqa Airport, Malta to Edinburgh Airport, whereupon the aircraft was refuelled and then returned to Malta, via Jersey, with more seamen. Later in March, two Dakotas carried over sixty school children from Hatfield to Paris, and during the spring and summer many passenger charters were flown to Paris carrying holidaymakers, mainly from London and Zurich. The Whitsun Holiday in May saw the operation of many charters to Paris from Northolt, Tollerton and Gatwick carrying tourists, while a party of students travelled from Gatwick to Frankfurt by Air Transport Charter Dakota.

By 1950, therefore, the scope of Air Transport Charter's operations was wide, and during the summer many holiday flights were made to European airports, including Amsterdam, Le Havre, Ostend, Paris, and Zurich. Many unusual charters were also undertaken - on June 2 1950 one of the Dakotas carried a cargo of pigeons from Gatwick to Nantes, and on October 6 the airline was responsible for carrying some of the members of the Halle Orchestra from London to Lisbon. By October, though, the airline had been taken over by Lambert Brothers, a firm of London based travel consultants, who, wishing to fully utilise their fleet of Dakotas during the winter months, moved Air Transport Charter's operations base from Jersey to Blackbushe in October 1950. During the winter months it was normally difficult to find work for the Dakotas in the Channel Islands, but the airline had obtained several contracts for flights from the London area. One of these contracts involved the carriage of missionaries from Amsterdam to various parts of Africa. In addition, Air Transport Charter often operated services on behalf of the State airlines - on November 15 one of the Dakotas left London bound for East Africa with B.O.A.C. cargo. Services were also flown from London to Milan and Rome carrying cargo under contract to B.E.A.C.

During 1951 Air Transport Charter operated many flights on behalf of B.E.A.C. and B.O.A.C., and Royal Air Force personnel flights were made to El Adem, Celle and Malta. Nightly newspaper flights were operated between Ringway Airport, Manchester and Nutts Corner, Belfast, but on March 27 1951 Dakota G-AJVZ crashed at Ringway while operating one of these flights, killing the crew of two. Many tourist flights were also flown on behalf of a variety of travel agents, with services connecting Blackbushe with Amsterdam, Dinard, Geneva, Innsbruck, Ostend and Paris. During 1950 Air Transport Charter had operated a series of twenty flights between Blackbushe and Calvi, on the Isle of Corsica, on behalf of Horizon Holidays. Over six hundred passengers were carried on these services during 1950, and on June 8 1951 one of the airline's Dakotas reopened this route for the summer. The summer of 1951 also saw the restoration of the flights from Blackbushe to Guernsey and Jersey, and the season proved to be a busy one for Air Transport Charter. Although the autumn brought a decline in the number of tourist flights operated by the airline, several Air Ministry and War Office contracts were secured to tide the company through the traditionally slack winter months. The services to the Channel Islands were, once again, suspended for the winter of 1951, and although they were reopened for the 1952 summer season, this was to be Air Transport Charter's last season of flying. The airline, for the third season in succession, flew Horizon Holidays' services to Calvi, and numerous tourist flights were made to European destinations, but during the summmer the company was prosecuted for the operation of an unlicensed scheduled service between Blackbushe and Jersey. Sadly, the charges were found to be true, and at the end of the season Air Transport Charter ran down its air charter operations, and ceased all flying on October 31 1952. The fleet of aircraft was sold, and Air Transport Charter now exists as an avionics firm based in Slough. At the time of Air Transport Charter's closure of air charter operations, the board of directors consisted of Messrs.A.M.Watson, F.C.Johnson, P.H.Renouf and Captain W.Gratrix, while Captain W.J.S.Wright acted as General Manager and Chief Pilot.

Fleet:

Miles M.57 Aerovan 4	G-AISE	c/n 6395	dd. 3.47,	sold 6.47.
	G-AISF	c/n 6396	dd. 3.47,	sold 6.47.
DH89A Dragon Rapide	G-AFFB	c/n 6409	dd. 6.47,	ISOC , 'St.Peter'.
	G-AGWC	c/n 6916	dd. 4.47,	sold 11.52, 'St.Lawrence'.
	G-AIUL	c/n 6837	dd. 5.47,	ISOC , 'St.Clement'.
Douglas C47A Dakota	G-AJBG	c/n 25448	dd. 4.47,	written off Bovingdon 20.5.48, 'St.Martin'.
	G-AJBH	c/n 25460	dd. 7.47,	ISOC , 'St.Helier'.

Douglas C47A Dakota	G-AJVZ	c/n	19361	dd. 6.48,	written off Ringway 27.3.51, 'St.Saviour'.
	G-AKIL	c/n	25282	dd.10.47,	sold 11.51, 'St.Mary'.
Douglas C47B Dakota	G-AKOZ	c/n	27188	dd. 9.50,	ISOC , 'St.Ouen'.

AIRVIEWS

Airviews Ltd.
Airviews (Manchester) Ltd.

Bases: Barton and Manchester (Ringway) Airports.

Airviews was formed early in 1946 with one Auster Autocrat. This aircraft, operating from Ringway and Barton Airports, was used for general ad-hoc charter work, aerial photography, aerial advertising and banner towing. In April 1950, this Autocrat was joined by an Autocar, but in September this new aircraft was lost in an accident near Leicester. Pleasure flying was also one of the company's major activities, and in September 1951 a Dragon Rapide was acquired. In 1953 the airline received a licence to operate a scheduled service from Manchester to the Isle of Wight, via Southampton, and this service was flown during the summer months by the Rapide. Financial problems resulted in a suspension of operations during 1957, but a Dove acquired during the year allowed the company to resume flying, and scheduled passenger services were opened from Manchester to Newcastle and Newquay. The service to the Isle of Wight via Southampton continued in operation, and by 1958 this Rapide service was flown three times a week between May and September. The services to Newcastle and Newquay were flown twice weekly, and thrice weekly respectively, by the Dove.

The airline specialised in providing services to racemeetings, and regular venues included Cheltenham, Doncaster, Leicester, Lincoln, Newbury, Newmarket, Ripon and York. In addition, Airviews had hoped to launch a daily Manchester to Teesside service, and a weekly service to Torquay, but all scheduled services were suspended during 1959, and the company put into liquidation. All the aircraft were sold, but later in the same year the company restarted operations as Airviews (Manchester) Ltd., and with an Auster the company continued operating pleasure flights, and carrying out aerial photography. These activities continue to this day with a Beagle Husky, and some of Airviews' clients include British Airways, British Railways, British Steel Corporation, Esso Petroleum, Ferranti, H.J.Heinz, I.C.I. Fibres, Kellogs, and many others. The pleasure flying is mainly carried out at Manchester Airport when the Husky is not required for photographic work.

Fleet:

Auster J/1 Autocrat	G-AGXN	c/n	1963	dd. 4.46,	sold 6.66, converted to J/1 Alpha 2.57.
Auster J/5B Autocar	G-AJYK	c/n	2908	dd. 4.50,	written off Leicester 18.9.50.
DH89A Dragon Rapide	G-AFRK	c/n	6441	dd. 5.56,	wfu. 3.59.
	G-AGDM	c/n	6584	dd. 9.51,	sold 3.57.
	G-AGSH	c/n	6884	dd. 5.56,	sold 4.57.
	G-ALBA	c/n	6821	dd. 9.55,	sold 2.59.
DH104 Dove 1B	G-AIWF	c/n	04023	dd. .57,	sold 7.59.
Auster J/5 Autocar	G-AMPV	c/n	2904	dd. .60,	written off Ringway 10.6.61.
Auster J/5P Autocar	G-AOBV	c/n	3171	dd. .66,	sold 1.71.
Beagle D5/180 Husky	G-ASNC	c/n	3678	rd. 9.70.	

AIRWAYS INDIVIDUAL RESERVATIONS

Airways Individual Reservations Ltd.

Base: Denham Airfield.

Airways Individual Reservations was an air charter operator formed at Denham Airfield in 1946. Operations started in July with a fleet consisting of Percival Proctors and a Q.6 - these aircraft were used on many charters to Europe carrying passengers. The Q.6 was even more adventurous, and many flights were made to North Africa and Italy in the hands of Captain A.R.Lewis, Airways Individual Reservations' Chief Pilot. However, while returning from one of these long range charter flights, this aircraft crashed at Jury's Gap, Dungeness when en-route from Le Bourget to Lympne in bad weather on Christmas Eve 1946. Although no one was seriously hurt, and the aircraft was not written off, the Q.6 was sold to Western Airways who collected the aircraft and rebuilt it at Weston. A Proctor was also sold in December 1946, and although operations continued into 1947 with the two remaining Proctors, operations ceased early in the year.

Fleet:

Percival Proctor 1	G-AHMX	c/n	K.260	dd. 7.46,	ISOC.
	G-AHUZ	c/n	K.302	dd. 8.46,	ISOC.
Percival Proctor 2A	G-AHMY	c/n	H.82	dd. 7.46,	sold 12.46.
Percival Q.6	G-AHOM	c/n	Q.42	dd. 7.46,	sold 12.46, after accident on 24.12.46.

AIRWORK

Airwork Ltd.
Transair Ltd.

Bases: Blackbushe, and later Gatwick Airports.

Airwork was formed in 1928, and initially developed Heston Aerodrome as an aviation and social centre. Flying instruction was undertaken from there, and later on elsewhere at other aerodromes including Perth and Shoreham. The flying school at Perth still exists today under Airwork ownership. The company also undertook aircraft maintenance, and its first air transport experience came in 1932 in Egypt, when the company helped to set up an airline which we know today as Egyptair. In the following year, Airwork assisted in the formation of Indian National Airways.

During the war, Airwork aircraft and their flying clubs were used to train Royal Air Force personnel, but once the war was over, Airwork returned to commercial aviation. In 1946, the company was appointed a dealer for the Airspeed Consul, and also acquired a fleet of Dragon Rapides and Vikings for use on passenger and freight charter and contract work. Airwork's first Viking entered service in April 1947, and by the end of the year, several of this type of aircraft were in service. The Vikings operated mainly passenger charters, and the airline also held contracts with several governments for the carriage of official passengers and their families. The company operated a Sudan government leave service, and this started during 1947. The aircraft left Wadi Halfa and Khartoum for London approximately twice a week, and on April 4 1950, Airwork carried its 10,000th passenger on this service. By this time, some three hundred and forty five round trips had been made carrying a total of 10,013 passengers, and 394 babes in arms. The Vikings also operated numerous other government contract services to the Middle and Far East, and also services within Africa, with its fleet of Vikings. During 1948 Airwork formed a subsidiary company in East Africa named Airwork (East Africa) Ltd., with its headquarters at Nairobi. This company developed air transport facilities in East Africa and provided an aircraft maintenance facility in the area. This company took over the aircraft interests in East Africa of the Uganda Company, and also acquired a shareholding in Noon and Pearce Ltd., of Nairobi. Airwork despatched two Bristol Freighters for these services, and initial operations were conducted in Tanganyika, where the aircraft were used to carry passengers and freight of all sorts including cars, spare parts, meats and fresh vegetables.

On November 10 1948, Airwork positioned one of its Bristol Freighters to Hamburg to start flying sorties on the Berlin Airlift. However, although a second Bristol Freighter joined the first aircraft at Hamburg, Airwork only flew 74 sorties in 219 hours of flying, and the airline was withdrawn from the airlift on February 12 1949. Two days later, on February 14, Airwork operated its first service on behalf of the War Office to fly wives and families of service personnel to the Middle East. These services were operated by Vikings, and flew from Blackbushe to either Cairo or Khartoum several times every week. During the next few years, Airwork's growth was based on this type of work until June 14 1952, Airwork Viking G-AJFS inaugurated the airline's newly approved 'Safari' service to Africa. Within two years, further services had been opened, and these colonial coach class services were being flown to Nairobi, Lagos and Salisbury. Using Vickers Vikings, these flights operated via Malta, Wadi Halfa, Juba, Khartoum and Entebbe.

In 1952, four Handley Page Hermes were purchased from B.O.A.C. for use on Ministry of Defence trooping flights to Kenya and the Egyptain Canal Zone. The first aircraft initially underwent a series of tropical trials and crew training with the airline, and the engines had to be specially converted to accept a different octane fuel. By 1952, Airwork was also operating a number of Dove in the Middle East on behalf of several oil companies including the Anglo Iranian Oil Company in Abadan, and the Iraq Petroleum Company in Haifa. During 1952, Airwork applied to operate a daily freight service, with Bristol Freighters, from Blackbushe to Aden via Port Sudan, but this service was never started. In 1954 the airline won an Air Ministry trooping contract to the Far East. This covered almost three years work, and provided for the carriage of some seven thousand troops each year in each direction between the United Kingdom and Singapore, and at the time it was the biggest single air trooping contract ever placed by the Air Ministry, and was worth some £1¼ million per year. Again the Hermes were used for this marathon service. At this time Airwork's shareholders included the Blue-Star Line and the Furness Withy Line - both large shipping companies. In 1954, a colonial coach class service was opened between London and Aden using Vikings, but in the following year Airwork was to make a change in its character of operations.

In September 1952, Airwork applied to the Air Transport Advisory Council for permission to operate an all freight service from London to Montreal and New York. In February 1953,the airline received approval to operate this service for ten years. However, the company had no long range all cargo aircraft capable of operating this service, and so the airline leased two Transocean Douglas DC4s to open these services. Therefore on Tuesday March 1 1955, a DC4 (still named 'Argentine Queen' from earlier days) left Heathrow Airport on its inaugural all freight service to North America. The DC4 flew via Manchester and Prestwick to collect further cargo, and made refuelling stops at Keflavik and Gander before arriving at New York's Idlewild Airport. Airwork Atlantic, as this service was known, had offices in London, Manchester, Birmingham and Glasgow, and in addition to the bi-weekly Monday service to New York, a bi-weekly Thursday service was flown to Montreal. Airwork soon realised that flying two leased veteran DC4s on these services across the Atlantic was not the most suitable way of operating these flights, and so Airwork placed an order with Douglas' for two all cargo DC6As. To connect with these trans-Atlantic freight services, Airwork opened connecting freight services to London using Vikings from Aden, Hamburg, Frankfurt and Zurich, but after little over nine months of operations, the trans-Atlantic freight services were suspended in December 1955, and the order for the DC6As was cancelled. The European freight services, which by now were serving Manchester, Birmingham, Glasgow, Frankfurt, Zurich, Hamburg and Dusseldorf continued in operation for a short while, but these too were suspended on March 8 1956. 1956 was, therefore, a year of recovery and consolidation for the company, and the fleet now mainly consisted of trooping Hermes. However, 1956 did see Airwork helping to form the new German airline Deutsche Flugdienst.

During 1958, Viscounts replaced the now ageing Vikings on the colonial coach class services to Africa which Airwork still operated in close association with Hunting-Clan, and later in the year Airwork took over another British independent - Transair. On Monday June 8 1959 Airwork opened the Blue Nile Viscount service between Gatwick and Khartoum on behalf of Sudan Airways. The Viscount 831 used on this service flew via Rome, Athens and Cairo, and on arrival at Khartoum on the Tuesday then operated two scheduled services between Khartoum and Cairo, before returning to Gatwick Airport on Thursdays. During 1959 the trooping work decreased, and many of the Hermes were sold or scrapped, while some of the Viscounts operated Transair services,and some of the colonial coach class services were transferred to Transair. Then, on July 1 1960, the company merged with its old associate Hunting-Clan, and another independent airline Air Charter, to form British United Airways.

Fleet:

Airspeed AS.65 Consul	G-AHEG	c/n	1052	dd. 4.46,	sold 4.48.	
	G-AIKR	c/n	4338	dd. 1.47,	sold 2.62,	used by Airwork Services Training
	G-AJNE	c/n	5139	dd.11.56,	wfu. 5.61,	used by Airwork Services Training
DH89A Dragon Rapide	G-AESR	c/n	6363	dd. 5.48,	sold 3.53.	
	G-AJDN	c/n	6860	dd. 4.47,	sold 6.58,	used by Airwork Services Training
	G-AKJS	c/n	W1002	dd.12.47,	sold 7.62,	used by Airwork Services Training
	G-AKRS	c/n	6952	dd. 4.48,	sold 7.62,	used by Airwork Services Training
	G-AKTZ	c/n	6482	dd. 5.48,	sold 4.53.	
Percival Proctor 1	G-AHMV	c/n	H.547	dd. 6.46,	sold 7.46.	
Percival Proctor 3	G-AIHD	c/n	H.171	dd. 6.53,	sold 6.58.	
	G-ALCK	c/n	H.536	dd. 7.54,	sold 55.	
	G-ALER	c/n	H.455	dd. 9.49,	wfu. 3.55.	
	G-AMBS	c/n	H.241	ca. 1.51,	sold 56.	
Percival Proctor 4	G-AKWL	c/n	H.682	ca. 7.48,	wfu. 9.56.	
Avro 652A Anson 1	G-AIRW	ex.	MH167	dd.10.48,	sold 7.50.	
	G-AKMV	ex.	EG239	dd. 6.48,	sold 1.51.	
Bristol 170 Freighter Mk.11A	G-AHJD	c/n	12736	dd. 8.46,	sold 7.47.	
Bristol 170 Freighter Mk.21	G-AHJD	c/n	12736	dd.10.48,	sold 4.52.	
	G-AICS	c/n	12762	lsd. 1.49,	ret. 10.49,	leased from Bristols.
	G-AICT	c/n	12763	lsd.10.47,	ret. 4.52,	leased from Bristols.
Vickers 498A Viking	G-AHOP	c/n	117	dd.10.52,	ISOM ,	trooping as WZ972.
Vickers 627 Viking	G-AIXR	c/n	233	dd. 4.47,	sold 5.59,	trooping as WZ355.
	G-AIXS	c/n	234	dd. 5.47,	written off	Blackbushe 15.8.54, trooping as WZ354.
	G-AJFP	c/n	235	dd. 6.47,	sold 12.48.	
	G-AJFR	c/n	236	dd. 7.47,	sold 4.58.	
	G-AJFS	c/n	237	dd. 9.47,	sold 2.58,	trooping as WZ311.
	G-AJFT	c/n	238	dd.10.47,	sold 3.60,	trooping as WZ306.
Vickers 634 Viking	G-AKTU	c/n	211	dd. 2.48,	sold 3.60,	trooping as WZ356.
	G-AKTV	c/n	208	dd. 2.48,	sold 5.59,	trooping as WZ357.
DH104 Dove 1	G-AHRJ	c/n	04004	dd. 6.50,	sold 10.50,	operated by Channel Islands Air Services.
	G-AKSS	c/n	04122	dd. 2.49,	sold 2.51,	operated by Channel Islands Air Services.
DH104 Dove 2	G-AMEI	c/n	04296	dd.10.50,	sold 6.51.	
Douglas C47A Dakota	G-AGIS	c/n	12017	dd.11.53,	sold 12.53.	
Douglas C47B Dakota	G-AGKC	c/n	25591	dd.10.49,	sold 2.52.	
Douglas C47A Dakota	G-AGYZ	c/n	12278	dd. 5.54,	sold 3.57.	
	G-AMBW	c/n	25275	dd. 6.50,	sold 11.53,	trooping as XB246.
Douglas C47B Dakota	G-AMRA	c/n	26735	dd. 4.52,	ISOM ,	operated by Transair.
	G-AMVA	c/n	33163	dd. 3.53,	sold .	
	G-AMZD	c/n	32860	dd. 3.53,	sold .	
	G-AMZW	c/n	27099	dd. .53,	sold 10.53,	never entered service.

G-ACIT is now the oldest British built transport aircraft still in flying condition in this country - here it appears in Air Navigation and Trading's colours at Blackpool on May 23 1954.

A M G Armstrong

August 26 1961 and Air Safaris' Hermes G-ALDA is prepared for another inclusive tour flight from Newcastle.

Ian MacFarlane

G-AHPC at Newcastle on June 3 1961 before operating an inclusive tour flight to Valencia, via Manchester.

Ian MacFarlane

Air Taxis was one of the many early post-war Proctor charter operators - G-AGYA being one of the aircraft used.

A J Jackson

Air Transport Charter's Dakotas operated numerous holiday flights during the summer months between Newcastle and Jersey with remarkable ninety per cent load factors, G-AKOZ being one of the aircraft used.

A J Jackson

Airviews, formed by Mr. John B Martin, became a limited company on December 1 1949, and operated many Rapides including G-AGDM illustrated here.

M G Armstrong

G-AFFB Air Transport Charter Dragon Rapide — A J Jackson

G-AIWF Airviews Dove — A J Jackson

G-ARGR Alidair Viscount 708 — Ian MacFarlane

G-ALPK Air Taxi Dragon Rapide — Ian MacFarlane

G-ARNB Anglian Air Charter Auster — T'portraits/John Goring

G-AYWF Air Taxis International Aztec — Tony Merton Jones

Douglas C47B Dakota	G-AMZX	c/n 26735	dd. .53, sold 9.53, never entered service	
Handley Page 81 Hermes 4	G-AKFP	c/n HP81/1	dd. 8.52, written off Calcutta 1.9.57, trooping as XD632	
	G-ALDA	c/n HP81/2	dd. .52, sold 10.59, trooping as WZ838.	
	G-ALDB	c/n HP81/3	dd. .52, written off Orleans, France 23.7.52, trooping as WZ839.	
	G-ALDC	c/n HP81/4	dd. .52, sold 6.59, trooping as WZ840.	
	G-ALDF	c/n HP81/7	dd. 2.52, written off Sicily 25.8.52, trooping as WZ841.	
	G-ALDG	c/n HP81/8	dd. 8.57, sold 10.59.	
	G-ALDO	c/n HP81/16	dd. 8.52, wfu. 3.59.	
Douglas C54A Skymaster	N4726V	c/n 10315	lsd. 3.55, ret. 12.55, leased from Transocean.	
	N5288N	c/n 10318	lsd. 3.55, ret. 12.55, leased from Transocean.	
Douglas DC6A	G-AOFX	c/n 44889	never delivered.	
	G-AOFY	c/n 45058	never delivered.	
Vickers 736 Viscount	G-AODG	c/n 077	dd. 1.58, ISOM.	
	G-AODH	c/n 078	dd. 1.58, ISOM.	
Vickers 831 Viscount	G-APND	c/n 402	dd. 2.59, ISOM.	
	G-APNE	c/n 403	dd. 3.59, ISOM.	

ALDERNEY

Alderney Air Charter Ltd.
Mossair Ltd.

Base: Alderney Airport.

Alderney Air Charter was formed as an air taxi and charter operator by Captain P.R.Moss in the summer of 1972. Based at Alderney Airport, operations started on September 13 1972 with one Cessna Skymaster which was initially used for many inter-Channel Islands flights, and for charters to airfields on the south coast of England and in north western France. During the spring of 1973 the Skymaster was busily engaged in carrying fruit, flowers and vegetables from Alderney to the mainland, and in July a Super Skymaster joined the fleet.

Early in 1975, Captain Moss formed an associate company named Mossair. This company acquired the former Rolls Royce Dove 7, and with this aircraft Mossair started air taxi and executive charter flying. After overhaul and respray by Fairflight at Biggin Hill, this aircraft entered service on April 8 1975 when it flew a charter from Biggin Hill, via Norwich, to Aberdeen, with the return flight operating via Edinburgh to Biggin Hill later the same day. The aircraft then flew a number of charters from Biggin on behalf of Fairflight, and also a flight on behalf of Volkswagen Spare Parts Ltd., with a service from Manston to Wassel, West Germany. The Dove, therefore, did not arrive in Alderney until April 23 1975, when it flew from Biggin Hill to Alderney. Since then, apart from services radiating from Alderney, the Dove has visited several other airports on charter flights, including Dinard, Liverpool, Ostend, Saarbruchen and Stavanger. The Dove was used at the 1975 International Air Fair at Biggin Hill during May for giving pleasure flights, and has since visited Biggin Hill on several occasions while under charter to Fairflight. When Fairflight are short of an aircraft, then the Mossair Dove is frequently hired, and has appeared on Fairflight's regular run from Southampton, via Newcastle, to Sumburgh.

Today, therefore, Mossair is kept busy operating a variety of passenger and cargo charter flights, while Alderney Air Charter now holds a contract for the carriage of lorry drivers between Hurn and Cherbourg. This service necessitates the operation of two return services daily, by Skymaster, and almost fully utilises one of Alderney Air Charter's aircraft. Other Skymasters, flown by Alderney Air Charter, are mainly used for passenger and cargo flights to Alderney, and during the summer months these aircraft are kept particularly busy flying people from England and north western France to Alderney. The transport of Channel Islands' flower and vegetable produce is also an important part of the company's activities during the spring and early summer.

Fleet:

Cessna 336	G-ATAH	c/n 0007	dd. 9.72.	
Cessna 337B Super Skymaster	G-AVJG	c/n 0715	dd. 7.73.	
	G-AVIX	c/n 0554	lsd.11.72, ret. 11.72,	leased from Skycab Europe.
	G-BBBL	c/n 0555	dd. 2.75.	
Cessna 337C Super Skymaster	G-BCBZ	c/n 0492	lsd. 8.74, ret. 9.74,	leased from General Aviation.
	G-BCBZ	c/n 0492	lsd. 6.75,	leased from General Aviation.
Cessna F150G	G-AVCU	c/n 0129	dd. 4.74.	
Piper PA27 Aztec 250D	G-AZSZ	c/n 27-4194	lsd.10.74, ret. 10.74,	leased from Express Aviation.
DH104 Dove 7XC	G-AJGT	c/n 04034	dd. 4.75,	operated by Mossair.

ALIDAIR

Alidair Ltd.
Alidair Cargo Ltd.

Bases: Hucknall, and later East Midlands Airports.

Alidair has, in just four years, made for itself a very firm place in the history of British air transport. Today, it is one of the few 'old fashioned' types of British airlines still left in existence. Alidair employs a staff of around eighty five people, with many members of the same family being employees - the wife of Alidair's managing director often flies as a member of the cabin staff. The personal touch within Alidair has led it to grow to an airline flying Viscounts on a very wide variety of services, and this all started from rather more humble beginings.

The Alida Packaging group purchased two Piper Twin Comanches in the summer of 1971 for use as executive transports for the company's directors. A Beagle Pup, was acquired in November, was used by the Alidair Flying Club to give flying instructions to members of the public and also to two of Alida's directors. A change of role came in January 1972 when Alidair Cargo was set up to operate cargo services from Hucknall. Captain Roger Dadd, a twenty nine year old pilot, became managing director of this new company, and in April he purchased the serviceable remains of the once proud Channel Airways Viscount 800 fleet. The fleet consisted of three 812s, and Alidair also bought a British Midland series 831 Viscount in May. After conversion, the British Midland and two of the Channel Airways aircraft entered service with Alidair, now based at Castle Donington due to the lack of customs facilities at Hucknall.

With this fleet of three serviceable Viscounts, Alidair operated its first charter on June 12 1972, when a party of schoolchildren were flown from Coventry to Graz. Numerous other passenger and many cargo charters were operated during the remainder of the year, and a subsidiary company, Alidair Holidays, arranged a large number of day trips from Castle Donington. These day trips served a number of destinations including Amsterdam, Brussels, Guernsey, Jersey, Ostend, Paris and Rotterdam. An all inclusive cost of around £17 included a free bottle of duty free whisky or gin! These trips proved popular enough for up to four flights a week to be operated at the height of the 1973 and 1974 summer seasons. One of the Viscounts was converted into a forty four seat VIP aircraft with a luxury interior, and this aircraft was originally intended for a series of aerial cruises around the Mediterranean and North Africa planned by Alidair Holidays for the winter of 1973, but the increase in the price of fuel and the general depression in the travel business, caused these plans to be postponed. This luxury Viscount was also intended for use by companies as a flying boardroom.

Late in 1972 an application was made to the Air Transport Licensing Board for permission to operate a scheduled passenger service from Castle Donington and Birmingham to Copenhagen and Malmo using Viscounts. This service was quickly approved, and on April 30 1973 the company's luxury Viscount inaugurated this thrice weekly service. Passengers flew on this service in first class comfort with a hot breakfast served on the outbound flight. However, rarely were more than twenty seats filled on these flights, and when the fuel crisis hit in December 1973, Alidair was forced to drop this service due to difficulties in obtaining fuel. During the spring and summer of 1973, several day trips were flown from Coventry, and on April 8 one of the Viscounts left for Southend on lease to British Air Ferries for the summer season. Also from Southend, Alidair flew charter and holiday flights during the 1973 summer season. These flights started on May 19 1973 with flights to Alicante and Palma on behalf of the Dee Tee Travel agency of Derby. During the summer other holiday flights were operated from Southend to Cologne, Cork, Ostend, Paris and Rotterdam, and all these services came to an end in October.

As a result of the fuel crisis, one of the Viscounts was sold to Arkia in March 1974, but in April Alidair's two remaining Viscounts started flying the springtime bulb charters for Clarksons. The Viscounts, starting at eight o'clock in the morning, flew continuously to Beauvais and Rotterdam from Birmingham, Castle Donington, Liverpool, Manchester, Newcastle and many other airports, finishing their task at ten o'clock every night. For the summer, day trips from Castle Donington and Coventry continued, and Alidair Holidays and other local travel agents chartered Alidair's Viscounts for inclusive tour services from Birmingham, Castle Donington and Manchester to many Mediterranean resorts, carrying up to seventy eight passengers on each flight. Also during 1974, many charters were flown from Castle Donington to Gothenburg in Sweden carrying spare parts for the Volvo Car Company. At its peak, this contract necessitated the operation of up to three flights a week. During the summer, the Viscounts also flew services on behalf of other airlines including Air Anglia on their services from Norwich, and Dan-Air on their Gatwick to Newcastle services. Alidair's variety in its operations was further displayed in July 1974 when it was awarded a contract to fly newspapers from Manchester to Inverness. Initially, these services were flown to Kinloss, and later to Lossiemouth, and by December 1974 a weekend flight on Saturday nights was also flown on behalf of the Irish Sunday Press for the uplift of Sunday newspapers from Dublin to Luton. This service continued for most of 1975, although Alidair transferred the English terminal on this service to Castle Donington, before this service was taken over by British Island Airways in October 1975.

In February 1975, Alidair once again explored a new field of charter operations when one of its Viscounts positioned to Aberdeen to start flying charters on behalf of oil companies in the area. This Viscount started flying services, on behalf of the Burmah Oil and Total Oil companies, from Aberdeen to Sumburgh, Bergen, Stavanger, Norwich and Amsterdam, and in addition a weekly flight was operated on behalf of Shell (Norway) from Aberdeen to Bergen. On April 2 1975, Alidair took over a twice nightly newspaper run from Liverpool to Belfast, but due to a lack of aircraft, this service was initially flown by Air Bridge Carriers. By May Alidair had solved its equipment problem with the acquisition of several Air Inter Viscount 700s, and these aircraft were operating the newspaper run to Belfast by early May. The Viscount 700 also proved ideal for the company's operations out of Sumburgh Airport - these aircraft could carry a good payload on a direct Sumburgh to Heathrow flight which Alidair often operates on behalf of an oil company.

While Alidair was building up its network of charter services in Scotland during the spring of 1975, the airline was also operating a large number of springtime holiday flights from British airports to the Continent. These services included charters from Castle Donington, Gatwick, Luton and Manchester to Beauvais and Rotterdam. These services ended in May, but day trips were still operated from Castle Donington to the Channel Islands, Paris, Amsterdam, Ostend and Brussels. For the summer season, Alidair operated a weekly inclusive tour holiday flight on Saturday afternoons from Castle Donington to Cork, but during the summer the airline concentrated on the expansion of its oil company services from Aberdeen and Sumburgh. With two Viscounts now based at Aberdeen, Alidair was now operating services on behalf of six oil companies. Most of these services radiated from Sumburgh and Aberdeen to Scandinavian destinations carrying oil rig crews, but a weekly service was flown on Thursdays during 1975 from Edinburgh to Dusseldorf and return, and many flights were operated from Norwich. Scandinavian destinations served on these oil rig support flights include Aalborg, Bergen, Esbjerg, Oslo and Stavanger, and a weekly service was also flown from Norwich to Pescara in Italy. This service, operated outbound on Wednesdays with the return flight on Thursdays, started during October 1975.

Many ad-hoc passenger and freight charters were also undertaken by Alidair during 1975, including several charters from Zurich to St.Mawgan carrying Swiss businessmen to view the China works in Cornwall as guests of English China Clays. An unusual ad-hoc charter on Tuesday July 29 1975 was the operation of a service from Northolt Airport to Le Bourget to celebrate the twenty fifth anniversary of a similar flight made by the prototype Viscount on July 29 1950. Alidair used their luxury Viscount (now G-AZNH) for this service, and it departed from Northolt with a party of VIPs aboard for Paris at exactly the same time as its predecessor twenty five years earlier. Another Alidair Viscount is now operated on the Dan-Air scheduled service between Lydd and Beauvais, while a Viscount is also due to be leased to Intra Airways from March 1976 onwards.

Alidair's work is, therefore, varied and interesting and their aircraft have become familiar sights at many airports, both at home and aboard. In addition to their oil rig support services, their newspaper flights, and their day trips to the Continent, Alidair also regularly carry British dealers for Volvo cars from Castle Donington to Gothenburg to view the Volvo Car works in Sweden. This variation of operations probably makes Alidair the most fascinating of all current British airlines.

Fleet:

Piper PA39 Twin Comanche C/R	G-AYZP	c/n	39-91	dd. 6.71,	sold 10.72.
Piper PA30 Twin Comanche	G-AVPS	c/n	30-1548	dd. 5.71,	sold 5.73.
Cessna 401A	G-AXBY	c/n	0032	rd. 9.72.	
Beagle Pup 150	G-AZFA	c/n	143	dd.11.71.	
Vickers 812 Viscount	G-ATUE	c/n	357	dd. 5.72,	never entered service.
	G-AVIW	c/n	358	dd. 5.72,	sold 1.75.
	G-AVJL	c/n	389	dd. 6.72,	sold 4.75.
	N501TL	c/n	353	dd.10.75.	
Vickers 831 Viscount	G-ASED	c/n	419	dd. 5.72,	sold 3.74.
Vickers 814 Viscount	G-AZNH	c/n	342	dd. 1.75,	damaged in heavy landing Guernsey 14.9.75.
Vickers 806 Viscount	G-AOYG	c/n	256	lsd. 2.75,	ret. 4.75, leased from British Airways.
Vickers 708 Viscount	G-ARBY	c/n	010	dd. 4.75,	operated initially as F-BOEC.
	G-ARGR	c/n	014	dd. 4.75,	operated initially as F-BOEB.
	G-ARIR	c/n	036	dd. 6.75.	
	G-BDIK	c/n	036	dd. 5.75,	operated initially as F-BGNT.
Vickers 724 Viscount	F-BMCG	c/n	052	dd.10.75.	

ALLIED

Allied Airways Ltd.
Aberdeen Airways Ltd.
E.L.Gander Dower Ltd.

Base: Aberdeen (Dyce) Airport.

On January 2 1934 Mr.E.L.Gander Dower registered a new airline company named Aberdeen Airways. Based at Aberdeen, Aberdeen Airways started flying with a de Havilland Dragon bi-plane and on July 28 1934 the company opened its own aerodrome at Dyce, which remains today as Aberdeen's airport. However, it was not until May 27 1935 that Aberdeen Airways inaugurated its first scheduled service - from Aberdeen to Orkney via Wick and Thurso.

During the summer, Aberdeen Airways operated an Aberdeen to Edinburgh service to connect with North Eastern Airways' Edinburgh to London service. While this was being operated, the service to Orkney was suspended, although it was resumed during December. Further expansion came on June 2 1936 when the airline opened a new service from Aberdeen to Shetland via Stromness. During 1937, Aberdeen Airways changed its name to Allied Airways, and in November Allied started carrying mail on the Aberdeen to Shetland service. When war broke out in 1939, Allied Airways continued flying services under a Government contract, and several Rapides were transferred to the company during the war years, and during the war Allied flew the following services - Aberdeen - Thurso - Kirkwall - Stromness; Thurso - St.Margaret's Hope - Stromness; Thurso to Kirkwall; and Aberdeen to Shetland. A further Rapide was added to the fleet in August 1946, and during the following winter Allied continued to operate its scheduled services from Aberdeen to Shetland, Wick and Orkney. When B.E.A.C. took over the scheduled services of most private airlines on February 1 1947, Allied was fortunate and managed to remain independent for several more weeks, but on April 12 1947 Allied handed over most of its surviving aircraft and scheduled services to B.E.A.C.

Mr.E.L.Gander Dower himself continued charter flying with three Rapides from Aberdeen under the title of 'E.L.Gander Dower'. For these operations he used G-ADAH, G-AIDL and G-AJGS, but these operations came to an end early in 1950.

Fleet: (Allied Airways Ltd.)

Short Scion	G-ACUV	c/n S.774	dd. 7.34,	sold 6.39.	
DH84 Dragon 1	G-ACAN	c/n 6000	bt. 9.34,	written off	Dunbeath 21.5.41, 'The Starling'.
DH84 Dragon 2	G-ACRH	c/n 6078	dd. 6.34,	written off	Dyce 13.7.34, 'Aberdonian'.
	G-ADFI	c/n 6100	dd. 9.35,	written off	Thurso 3.7.37, 'The Silver Ghost'.
	G-ACNJ	c/n 6072	dd. .35,	wfu. .46,	'Sir Rowland'.
DH86B	G-AETM	c/n 2353	dd. 7.37,	sold 4.39,	'Silver Star', later 'The Norseman'.
DH89 Dragon Rapide	G-ADDE	c/n 6282	dd. 6.35,	sold 11.37,	'Aberdonian'.
	G-ADDF	c/n 6284	dd. 8.35,	sold 9.35.	
	G-ACZE	c/n 6264	dd. 7.42,	written off	Orkney 27.12.45, 'The Don'.
	G-ACZF	c/n 6268	dd. 6.39,	ISOT ,	'Carina'.
	G-ADAH	c/n 6278	dd.10.38,	ISOT ,	'Pioneer'.
DH89A Dragon Rapide	G-AGDM	c/n 6584	dd.11.41,	ISOT ,	'Eldorado'.
	G-AGHI	c/n 6455	dd. 3.44,	ISOT ,	'The Shetlander'.
	G-AIDL	c/n 6988	dd. 8.46,	ISOT ,	'The Wanderer'.
DH84 Dragon 1	G-ACLE	c/n 6044	bt.12.37,	sold 9.39,	'Old Bill'.

Fleet: (E.L.Gander Dower)

DH89 Dragon Rapide	G-ADAH	c/n 6278	ISOF ,	ISOC.	
DH89A Dragon Rapide	G-AIDL	c/n 6968	ISOF ,	ISOC.	
	G-AJGS	c/n W.1001	dd. 3.48,	ISOC,	rebuild of G-ACZE.

ANGLIAN

Anglian Air Charter Ltd. Base: North Denes Airfield, Great Yarmouth.

Anglian Air Charter was formed in 1950 by Mr. and Mrs.L.G.Wright to operate pleasure flights in the Great Yarmouth area. With an Auster, Anglian Air Charter started operations from their own airfield at North Denes, near Great Yarmouth. This grass airfield was built by Mr. Wright, and was capable of accepting most types of light aircraft provided they could land and take off in one thousand eight hundred feet of runway. During the summer months the company took hundreds of enthusiastic flyers aloft in their fleet of Auster aircraft, while during the winter Anglian Air Charter existed by carrying out aerial photographic work, and by operating the occasional charter flight.

Anglian Air Charter's Auster also became familiar sights at many air displays during the summer months in East Anglia. In 1969, two Cessnas were acquired to replace the now ageing Austers, and with the start of oil exploration in the North Sea, an Islander was purchased in March 1970. Anglian Air Charter had already assisted in the formation of Rig-Air, and during the summer of 1970 this Islander was used both for pleasure flying from North Denes, and also for charter work associated with the search for oil in the North Sea. With further expansion in mind, Anglian Air Charter announced on July 15 1970 that at the end of the month it would merge with Rig-Air and Norfolk Airways to form a new airline named Air Anglia. Thus, on August 1 1970 the name of Anglian Air Charter disappeared,while the name of Air Anglia rose very quickly to become one of Britain's best known airlines. Anglian Air Charter's work at North Denes continues today under the Air Anglia banner, and a new subsidiary company named North Denes Aerodrome Ltd., has been formed to look after the operation of the aerodrome.

Fleet:

Auster 5	G-AKOT	c/n 1469	dd. .56,	written off	Gorleston, Norfolk 9.9.62.
	G-AKPI	c/n 1088	dd. .51,	sold .61.	
Auster J/1N	G-AJEB	c/n 2325	lsd. .68,	ret. 8.68,	leased from Chrisair.
	G-APTU	c/n 3412	dd. 9.60,	sold 3.65.	
Auster J/5G Autocar	G-ARNB	c/n 3169	rd. 5.64,	sold . .	
	G-ASFK	c/n 3276	dd. 5.63,	ISOM.	
Cessna 170B	G-AWOU	c/n 25829	dd. .69,	ISOM.	
Cessna F172H	G-AVEC	c/n 0405	dd. 5.69,	ISOM.	
Cessna 206A	G-ATLT	c/n 0523	dd. 4.70,	ISOM.	
BN2A Islander	G-AXVP	c/n 127	dd. 3.70,	ISOM.	

ANGLO CONTINENTAL

Anglo Continental Air Services Ltd. Base: Burford Aerodrome.

Although not one of Britain's major post war charter operators, Anglo Continental Air Services, nevertheless, deserves its place as one of the many post war charter operators. This company was registered as an airline operator on June 24 1946, and operations started during the same month with two Austers flying ad-hoc charters from Burford Aerodrome in Oxfordshire. In addition, the company undertook a small amount of pleasure flying, and although a third Auster joined the fleet in 1949, the airline never undertook a very large amount of charter work, and all operations had ceased by the end of 1950.

Fleet:

Auster J/1 Autocrat	G-AGTW	c/n 1839	dd. 6.46,	ISOC.
	G-AHHH	c/n 2011	dd. 6.46,	sold 11.50.
Auster 5	G-AJVT	c/n 1495	rd.11.49,	ISOC.

ANGLO NORMANDY

Anglo Normandy Airways Ltd.
Anglo Normandy Aviation Ltd.

Base: Guernsey Airport, Channel Islands.

Anglo Normandy Airways was an air taxi and executive charter operator formed on January 9 1973 by Captain N.Stevens and Captain J.Smith - both people were formerly pilots with Sagittair. Initially, air taxi operations were conducted with a leased Piper Apache, and the company's first flight took place on January 12 1973 with this aircraft. Based at Guernsey Airport, Anglo Normandy set up a maintenance base at this Channel Island airport in March, and during the same month acquired its own aircraft, a second hand Piper Aztec. This aircraft arrived in Guernsey on March 31, but did not operate its first service until June 5 1973. The leased Apache was returned to its owners in March, and to allow the company to continue flying until its own Aztec entered service, two Aztecs were leased from Spooner Aviation. From Guernsey Anglo Normandy undertook many charter flights to southern England and north western France, and with a rapid build up in business further Aztecs were also acquired. However, in August 1974 one of Anglo Normandy's founders, Captain N.Stevens, was killed in New Zealand while in the process of ferrying a new Navajo Chieftain. This caused the company to suspend all operations, and the aircraft were either stored or sold.

On January 9 1975 the company was reformed as Anglo Normandy Aviation by Messrs.J.S.Hamilton and Thomas R. Miller, and operations were restarted with a fleet of two Piper Aztecs. These aircraft are now flown on numerous air taxi flights from the Channel Islands to many destinations, including Bournemouth, Cherbourg, Dinard, Exeter, Gatwick, Quimper, Southampton and St.Brieuc.

Fleet:

Piper PA23 Apache 160	G-APBD	c/n 23-1781	lsd. 1.73, ret. 3.73,	leased from Channel Aviation.
Piper PA27 Aztec 250C	G-AYWF	c/n 27-3911	lsd. 4.73, ret. 4.73,	leased from Spooner Aviation.
	G-BAAJ	c/n 27-2569	lsd.11.73, ret. 1.74,	leased from Spooner Aviation.
	G-BAWB	c/n 27-2783	dd. 3.73, sold 5.74.	
	G-BCBM	c/n 27-3006	dd. 7.74, sold 9.74.	
Piper PA27 Aztec 250D	G-BADE	c/n 27-4205	lsd. 4.73, ret. 4.73,	leased from Spooner Aviation.
Piper PA27 Aztec 250E	G-BCDS	c/n 27-5283	dd. 4.74.	
	G-BCKJ	c/n 27-5376	dd. 8.74, sold 4.75,	never entered service.
	G-BCJR	c/n 27-5378	dd. 6.74.	

AQUILA

Aquila Airways Ltd.

Base: Hamble Water.

Aquila Airways was formed on May 18 1948 by Mr.Barry T.Aikman, D.F.C., formerly the General Manager of the Lancashire Aircraft Corporation, and Chief Navigation Officer of Transport Command during the War. In 1939 he had joined No.210 Flying Boat Squadron in the Royal Air Force, and it was, therefore, natural that he should establish an airline operating a fleet of large flying boats. In July 1948 two Sunderlands were purchased from B.O.A.C., and after conversion these aircraft entered service on the Berlin Airlift.

On August 4 1948 these two aircraft (G-AGEU and G-AGIA) started flying sorties on the airlift from Finkenwerder, on the Elbe, to Havel Lake on the outskirts of Berlin. These aircraft were ideal for transporting bulky commodities, and cargoes included salt, flour and meat, but with the prospect of the formation of ice on the Havel Lake with winter approaching, Aquila Airways was withdrawn from the airlift. Aquila was withdrawn on December 15 1948, by which time they had flown a total of two hundred and sixty five sorties in just over seven hundred flying hours. By the end of the airlift, Aquila was flying three Sunderlands, and during the winter months more aircraft were obtained from B.O.A.C., and either stripped for spares or made serviceable at Aquila's base on Hamble Water. By the end of February 1949, Aquila had purchased twelve B.O.A.C. Sunderlands - the aircraft were delivered to Aquila as soon as B.O.A.C. withdrew them from service. The price paid for this fleet was not disclosed at the time, but it was believed to be well below the real aircrafts' value.

Once the airlift had ended, Aquila looked around for other work, and although they had originally hoped to operate world wide ad-hoc charters, the airline had to shelve these ideas after witnessing the collapse of so many other British airlines engaged in this activity. Thus, Aquila applied for permission to operate, under a B.E.A. Associate Agreement, a scheduled passenger service from Southampton to Madeira. This service was quickly approved, and on March 24 1949, Aquila's Chief Pilot, Captain D.Pearson, flew the Sunderland G-AGEU on a proving flight over this route. On this flight, Aquila carried a crew of eight, one hundred pounds of Royal Mail and a party of guests of Aquila Airways. All went well on the proving flight, and two months later on May 14 G-AGEU inaugurated the first scheduled service between Southampton Water and Madeira. This service was then flown once weekly throughout the summer, and before returning to Southampton from Madeira, Aquila's Sunderland would operate a service from Madeira to Lisbon and return. Although the service to Madeira from Southampton proved to be extremely popular, Aquila had built up a large fleet of former B.O.A.C. Sunderlands, and it needed to find further work for all these aircraft. Therefore, Aquila undertook various charters to many unusual destinations.

In June, one of Aquila's Sunderlands operated the first of many flights from Falmouth to the Isles of Scilly carrying holidaymakers. Other charters undertaken during June included the carriage of forty two crew members of a tanker, from Aden to the British Isles. The aircraft flew via Alexandria, Malta and Marseilles on this trip, and later in the same month the company flew a charter from Hull to Helsinki. In Helsinki, Aquila collected twenty seven Finnish seamen and flew them to Hull on June 19 - not only was Aquila's Sunderland the largest flying boat ever to land in Finland, but it was also the first commercial flying boat to land on the Humber for over ten years. In August, two Sunderlands carried a party of seventy four travel agents from Southampton to Le Havre, and by the end of the season business was very buoyant. However, the utilisation of the aircraft was still not as high as Mr.Aikman would have liked - during 1949 Aquila had hoped to operate a series of aerial cruises around the Mediterranean. These cruises, scheduled to start on June 15 1949, would have been operated on behalf of Thomas Cook, and the five thousand mile journey would have taken in Bordeaux, Cyprus, Malta, Marseilles, Rome and Venice in its itinerary.Sadly, though, these cruises were never flown due to lack of support.

The scheduled service to Madeira had carried almost three thousand passengers throughout the season, and with the success of this service, Aquila Airways were encouraged to look into the possibility of opening further services. Early in 1950 the airline announced its intentions to open scheduled services from Falmouth to Edinburgh (Leith) and Glasgow, with an optional stop at Southampton. The company took bookings for this service during the spring and early summer, hoping to start scheduled services over this route on June 1 1950. Later, it was announced that a limited service from Southampton only to Edinburgh and Glasgow would commence on July 4 1950. Bookings, though, were extremely low, and although G-AGJN operated a proving flight from Southampton to Edinburgh and Glasgow on June 1, the service was never opened.

The apathy shown by the public to Aquila's aerial cruises and services to Scotland was, fortunately, not displayed on the services to Madeira which, by 1950, had become very popular. Aquila's large fleet of aircraft also operated a variety of other services, and the granting of a B.E.A. Associate Agreement in 1950 allowed Aquila to open a scheduled passenger service between Southampton and Jersey. This service was inaugurated by Aquila on July 7 1950, and the arrival in St.Aubin's Bay, Jersey, of Aquila's Sunderland caused quite a stir amongst the islanders. This service was flown twice every Saturday throughout the summer, and continued until the end of the 1951 season. Aquila's Sunderlands were also chartered regularly by shipping companies to transport their employees around the world - one such charter operated by Aquila early in July 1950 was for the carriage of four employees from Southampton to Aden.

By 1951 Aquila was well known as both a scheduled and charter airline, and with the demise of most flying boat operators, it was comforting to watch the success of Aquila's operations. During the summer of 1951, services were again operated from Southampton Water to Jersey and Madeira, in addition to an increasing amount of charter work. In October 1951 the company's Sunderlands flew several charter flights from Southampton Water to Gibraltar with seamen, and these flights continued for much of the winter. During the winter of 1951 Aquila introduced its first Solent aircraft - this type had a considerably higher cruising speed than the older Sunderlands. The standard of comfort of the Solents was also superior, and the first aircraft (G-AKNU) was laid out as a forty one seater with a bar area, and two decks which were connected by a spiral staircase.

In 1952 the service from Southampton to Madeira was extended to Las Palmas in the Canary Islands - the run to Las Palmas was now operated by the Solent, while the shuttle service from Lisbon to Madeira was flown by the Sunderlands. During the early part of the year the company flew several more charter flights to Gibraltar with seamen, and in April Aquila carried out a survey of the route between the Falkland Islands and Montevideo on behalf of the Falkland Islands Company. Other charter work carried out by Aquila during 1952 included the transport of many of the competitors in the 1952 Olympic Games from the British Isles to Helsinki. The early part of 1953 sadly saw the loss of two Sunderlands in accidents, and in March Aquila Airways was taken over by the British Aviation Services Group. However, this take-over caused no major changes within Aquila, although later Mr.Aikman did resign from his position as Managing Director, and he took no futher part in the running of the airline. No new routes were opened in 1953, although Aquila did find an extra source of revenue by operating a series of one hour aerial cruises from Southampton, carrying passengers who wished to view the spectacle of a large part of the Royal Navy lined up off Spithead for inspection by Her Majesty the Queen, during her Coronation year. During 1953 Aquila carried six thousand passengers on its scheduled services, and during September, with further expansion in mind, Aquila opened negotiations with the Ministry of Supply for the purchase of the three Saro Princess flying boats. These most graceful of flying boats, origininally built for B.O.A.C., had been placed into storage at Calshot pending a decision on their fate. Sadly, though, Aquila failed to come to any form of agreement with the Ministry, and the airline was forced to buy further Solents.

By 1954, the service to Madeira was flown twice weekly, while the service to the Canary Islands was flown weekly. On June 3 1954 Solent G-ANAJ opened a direct scheduled service between Southampton and the Isle of Capri. This was the first ever direct air link between the British Isles and Capri - the flight took eight and three quarter hours including a one and a half hour stop at Marseilles. The service was flown weekly during the 1954 season, leaving Southampton every Thursday morning, while the return flight left Capri on the following morning. This new service helped to raise Aquila's passenger figures for 1954 to over eight thousand passengers. In October 1954 Aquila applied to the Air Transport Licensing Board for permission to operate an all freight scheduled flying boat service from Southampton to Perth, Western Australia, with stops at Malta, Cyprus, Bahrein, Kuwait, Karachi, Ceylon, Singapore and Sourabaya. This service, though, was never started.

With the continuing encouraging results of Aquila's scheduled services, the airline announced the introduction of further new scheduled services in 1955. On June 4 a once weekly flight to Santa Margherita from Southampton was started, while a once weekly flight to nearby Genoa was started on the following day, June 5 1955. Both these services left Southampton in the very early hours of the morning - at three o'clock - to enable passengers to arrive at their destinations in time for breakfast, thus allowing them the best part of an extra day on holiday. Both these services ran throughout the summer, but they came to an end on September 25. The service to Capri was only flown fortnightly during June and July 1955 - the first schedule being flown on June 2. However, during August and September, with increased passenger loads, this service was flown weekly.

Two new holiday routes had, thus, been introduced by Aquila Airways during 1955, and on January 8 1956 a new Solent (G-AOBL) opened a direct service from Southampton to Las Palmas. During June the airline opened another new service when Montreux, on Lake Geneva, was added to Aquila's route network. Prior to this, in May 1956, Mr.Barry Aikman resigned from the Board of directors, and the year was also marked by several other setbacks. The scheduled service to Capri was suspended, and in September one of the Solents was lost when it was blown from its moorings during a gale at Santa Margherita on the night of September 25 1956. Although a replacement aircraft was soon found, Aquila's operations were gradually being run down. On November 15 1957 one of the Solents crashed into a hill on Chessel Down, near Shalcombe on the Isle of Wight, and thirty five people lost their lives. Although operations continued during 1958, it was announced during the year that Aquila Airways would close down at the end of the season. The lack of a suitable replacement for the flying boats coupled with the increase in the amount of charter airlines flying holidaymakers to their Mediterranean playgrounds, caused Aquila to abandon any further expansion plans, and to close down.

Therefore, on September 30 1958 Aquila Airways ceased all flying. The three surviving Solents were ferried to the Tagus estuary near Lisbon, where it was hoped that a new airline could be started using these three aircraft. Sadly, though, this was not to be, and after lying in Lisbon for almost thirteen years with decaying fabric on the control surfaces, and corrosion in the hulls, the three Solents were finally dismantled. Aquila Airways was the last operator of large flying boats in the British Isles, and it will be remembered, by many, for a long time to come as one of the most remarkable of all post war British airlines.

Fleet:

Short S25 Sunderland 3	G-AGER	ex.	JM660	dd.12.48,	wfu. 7.56,	'Hadfield'.
	G-AGEU	ex.	JM663	dd. 7.48,	wfu. 3.53,	'Hampshire'.
	G-AGHZ	ex.	ML727	dd. 1.49,	wfu. 1.49,	'Hawkesbury'.
	G-AGIA	ex.	ML728	dd. 7.48,	wfu. 2.51,	'Hazlemere'.
	G-AGJJ	ex.	ML751	bt. 2.49,	wfu. 1.51,	'Henley'.
	G-AGJK	ex.	ML752	bt. 2.49,	wfu. 1.52,	'Howard'.
	G-AGJL	ex.	ML753	bt. 2.49,	wfu. 1.52,	'Hobart'.
	G-AGJM	ex.	ML754	dd. 2.49,	wfu. 1.52,	'Hythe'.
	G-AGJN	ex.	ML755	dd. 2.49,	written off	Funchal Bay 21.1.53, 'Hudson'.
	G-AGKY	ex.	ML789	bt. 1.49,	wfu. 5.53,	after take off accident at Southampton, 'Hungerford'.
	G-AGLA	ex.	ML791	dd. 1.49,	wfu. 8.49,	'Hudson'.
	G-AHEO	ex.	JM716	dd. 7.48,	wfu. 11.49,	'Halstead'.
	G-AHER	ex.	PP142	bt.10.49,	wfu. 1.52,	'Hemsdale'.
Short S25 Sandringham 1	G-AGKX	ex.	ML788	rd. 5.49,	wfu. 3.53,	'Himalaya'.
Short S45 Solent 2	G-AHIN	c/n	S.1302	dd. .56,	ISOC.	
Short S45 Solent 3	G-AKNU	c/n	S.1299	dd.12.51,	written off	Shalcombe 15.11.57, 'Sydney'.
	G-ANAJ	c/n	S.1293	dd. 4.54,	wfu. 9.56,	'City of Funchal'.
Short S45 Solent 4	G-ANYI	c/n	SH.1558	dd. 1.55,	ISOC ,	'Awateri'.
	G-AOBL	c/n	SH.1556	dd. 5.55,	ISOC ,	'Aotearoa'.

ARROW

Arrow Air Services (Charter) Ltd.
Arrow Air Services (Engineering) Ltd.

Base: Shipdham Airfield, Thetford.

Arrow Air Services was formed at Shipdham Airfield, near Thetford in Norfolk in September 1969. The founders of the company were former racing driver Mr.Jack Sears, Mr.Nigel Wright and Captain Robert Brook. Two seperate companies were formed - Arrow Air Services (Charter) to undertake air taxi and general air charter work, and Arrow Air Services (Engineering) to undertake aircraft maintenance and overhaul work. Initially, the company cleared the airfield at Shipdham, carrying out resurfacing work to the main runway and taxiways and constructing hangarage facilities. This completed, a Twin Comanche was acquired in May 1970, and this started air taxi and executive charter operations on June 16. During the following month, the airline's Mooney flew to Castle Donington to collect a Dragon Rapide, which later arrived at Shipdham, but was never used by Arrow. Since 1970, the company has gradually expanded and has been responsible for the maintnenance of many vintage aircraft including Austers, Dragon Rapides and Tiger Moths. The air taxi side of the business has also been gradually built up, but this progress has not been without its setbacks. The Twin Comanche acquired in May 1970, was damaged later in the year and was out of service for several months while being repaired. During its absence many aircraft were leased, including locally based Twin Comanches and Senecas, and then on January 23 1973 G-AXRW captained by Captain Bob Snook (managing director and chief pilot of the company) crashed near Shipdham while inbound from Aberdeen, tragically killing five of the six people on board. Despite this setback, Arrow has continued to expand, and although most of its charters are flown from Shipdham, the Twin Comanches also regularly fly charters from Norwich, where customs facilities are available for Arrow's international flights. The airline makes regular trips to Amsterdam, and also to Aberdeen in connection with oil rig support work.

Fleet:

DH89A Dragon Rapide	G-AKRS	c/n 6952	dd. 7.70,	sold 11.70.
Mooney M.21	G-ATHW	c/n 805	dd. 6.70.	
Piper PA30 Twin Comanche	G-ATYF	c/n 30-1205	dd.11.72,	, 'Brunflug'.
	G-AXRW	c/n 30-774	dd. 5.70,	written off Shipdham 24.1.73.
	G-AYAF	c/n 30-2000	dd. 5.73.	
Piper PA34 Seneca	G-AZON	c/n 34-50081	lsd. 72,	ret. .72.
Piper PA27 Aztec 250A	G-ASHH	c/n 27-63	lsd. 3.75.	

ASSISTAIR

Assistair Ltd.

Base: Glasgow Airport.

Assistair was formed in December 1974 at Glasgow Airport by former Airgo captain Geoff Rosenbloom. A base was also set up at Aberdeen to take full advantage of the very lucrative oil rig support business. Two Aztecs were purchased, and services started during January 1975. In addition to the services operated from Aberdeen and Glasgow, services were also regularly flown from Sumburgh and Edinburgh, and many flights were operated to continental destinations including Brussels, Amsterdam, Rotterdam, Stavanger, Bergen and Esbjerg. In addition, Assistair held a contract with the Post Office for the operation of mail flights from Edinburgh to Aberdeen and Montrose. The company operated these services during the night between Monday and Saturday using its Aztecs, but in 1975 the company sold out to the Glasgow based Burnthills Aviation - the name of Assistair disappeared, and Geoff Rosenbloom left to form another air taxi company.

Fleet:

Piper PA27 Aztec 250E	G-BAVW	c/n 27-4797	dd. 1.75,	ISOT.
	G-BBCM	c/n 27-4634	dd. 1.75,	ISOT.

ASTRAL

Astral Aviation Ltd.
Northern Air Charter Ltd.

Bases: West Hartlepool, later Newcastle Airport.

Northern Air Charter was formed at West Hartlepool Airport in January 1946, and initially flew one Percival Proctor V on air taxi and light freight work. In addition, the company was responsible for the operation of West Hartlepool Airport, including the supply of fuel and other essential services. In 1947 two Airspeed Consuls were added, and during the year these aircraft flew passenger charter services from Newcastle via Carlisle to the Isle of Man.

On August 27 1947, another company named Astral Aviation was registered as an airline company by Arthur C. L.Barrett and Norman Wilkinson - both Newcastle businessmen. This company took over the operations of Northern Air Charter in 1948, and a Dragon Rapide was added to the fleet. A base was maintained at Newcastle Airport, and a service to the Isle of Man was again flown during the 1948 summer season. This time the service was flown on a scheduled basis under a B.E.A. Associate Agreement. The company also undertook several charter flights to airports on the Continent, but at the end of the year Astral Aviation was taken over by the Lancashire Aircraft Corporation.

Fleet:

Percival Proctor V	G-AGTB	c/n Ae.8	dd. 1.46,	sold 1.46.
	G-AGTD	c/n Ae.9	dd. 2.46,	written off Carlisle 18.10.48.
	G-AHTD	c/n Ae.60	dd. 7.46,	written off Flushing 22.12.47.
	G-AHTK	c/n Ae.68	dd. 8.46,	ISOT.
Airspeed AS65 Consul	G-AJGA	c/n 5117	dd. 5.47,	ISOT.
	G-AJLH	c/n 5126	dd. 6.47,	ISOT.
DH89A Dragon Rapide	G-AKNN	c/n 6598	dd. .48,	ISOT.
Auster J/1 Autocrat	G-AJRC	c/n 2601	dd. .48,	ISOT.

ATLAS

Atlas Aviation Ltd.
Guernsey Air Charter Ltd.

Bases: Heston, later Elstree Aerodromes.

Atlas Aviation was registered as an airline company by Mr.G.J.Dawson on October 5 1946. However,, prior to this the company had started ad-hoc passenger and freight charters from Heston in July 1946 with a Consul and a Proctor. By October, most of the operations were being conducted from Elstree, and later in the year Atlas formed an associate company in Guernsey named Guernsey Air Charter. One Proctor and two Consuls were transferred to this company for inter-island flights, and also for use on charters from the Channel Islands to airports on the south coast of England during the spring and summer of 1947. The Proctor, though, was lost in the Channel in April 1947 while en-route from Guernsey to Lympne with passengers. The aircraft's engine had stopped while overthe sea, and the aircraft was forced to ditch; this led to an inquiry, during which it could not be established who actually owned the aircraft. However, flying continued during the early part of the summer, and a Miles Gemini was purchased in July, but it was shortly after this that Atlas ceased all operations.

Fleet:

Percival Proctor V	G-AHWP	c/n Ae.69	dd. 7.46	ISOC.
	G-AHWZ	c/n Ae.82	dd. 9.46,	sold 5.47.
	G-AHZX	c/n Ae.83	dd.10.46,	written off English Channel 16.4.47.

Airwork operated a large fleet of Vikings during the early post-war years mainly on Air Ministry and War Office contracts, and G-AIXR was one of the many aircraft used.

Vickers Ltd.

G-ALDF is seen here taxying out at Blackbushe for a crew training detail shortly after its delivery to Airwork on February 27 1952. For this crew training, several BOAC personnel were seconded to Airwork for the spring.

Flight

Another of Airwork's Hermes, G-AKFP, is pictured here on the apron at Blackbushe Airport on September 11 1954.

R A Scholefield

G-AJAX Guernsey Air Charter Consul — A J Jackson

G-AOBL Aquila Airways Solent — J G Lambert

G-BCJR Anglo-Normandy Aztec E — Ian MacFarlane

G-BBCM Assistair Aztec E — Stan Davidson

G-AXWR Aurigny Islander — Austin J Brown

G-AZJA Aurigny Trislander — Austin J Brown

"Taxi Sir ?". Two of Airswift's fleet of four Senecas await customers at their Blackbushe base.

Tony Merton Jones

Mossair's colourful Dove at Biggin Hill shortly after it had operated its first charter flight early in April 1975.

Tony Merton Jones

Allied Airways was one of the many British airlines whose scheduled services were taken over by BEA in 1947. Here G-AIDL is seen shortly after roll-out at Witney Aerodrome in 1946.

A J Jackson

Aquila's Short Solent G-ANAJ at rest at Funchal.

Flight

Birkett Air Service operated several Rapides on charter flights from Croydon between 1947 and 1953.

A J Jackson

Blue-Line Airways' only Dakota G-AGNG spent most of its short career with the airline flying charters from Bovingdon.

A J Jackson

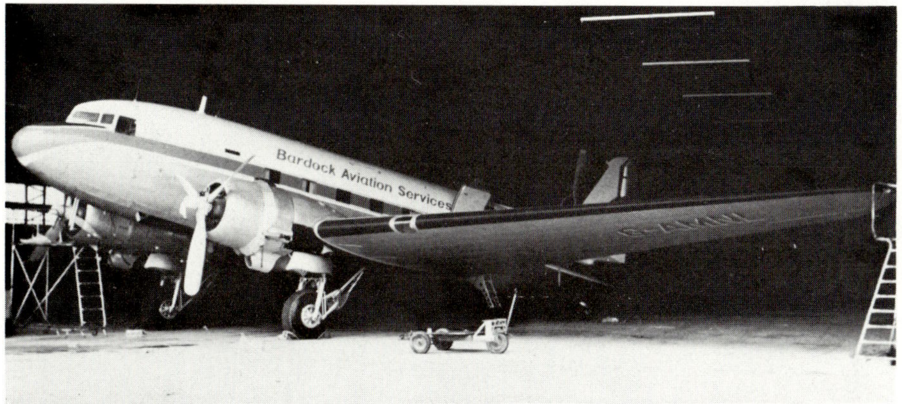

G-AMNL Bardock Douglas DC-3 via Stan Wilkinson

G-ATJZ Baylee Air Charter Aztec Ian MacFarlane

G-ADVJ Bond Air Services DH 86B Aviation Photo News

G-AYGZ Baron Air Charter Baron 58 Ron Cranham

G-BBIM Blue-Air Cessna 310 Ian MacFarlane

G-BBYK Bristol Air Taxis Aztec Ian MacFarlane

Euravia's first Constellation was G-ARVP, and it is seen here at Newcastle on Jnue 16 1965.

Ian MacFarlane

Britannia Airways' tragically famous Britannia G-ANBB is pictured here at Newcastle on June 16 1965.

Ian MacFarlane

The first Boeing 737 for Britannia Airways was G-AVRL, and it is seen here shortly after take-off from Luton Airport one summer's day in 1974.

Tony Merton Jones

Airspeed AS65 Consul	G-AHJY	c/n 2647	dd. 2.47,	ISOC.
	G-AHMA	c/n 3428	dd. 7.46,	written off Lyons, France 24.12.46.
	G-AIDW	c/n 2956	dd. 9.46,	ISOC, operated for Guernsey Air Charter.
	G-AJAX	c/n 3331	dd. 4.47,	ISOC, operated for Guernsey Air Charter.
Miles M.65 Gemini 1A	G-AJWF	c/n 6291	dd. 7.47,	ISOC.

AURIGNY

Aurigny Air Services Ltd. Bases: Alderney, Hurn and Staverton Airports.

This now famous third level operator was formed as a subsidiary of Glos-Air of Staverton late in 1967. Aurigny Air Services then apllied to the A.T.L.B. for permission to operate scheduled inter-Channel Island passenger flights with a fleet of Islanders. Approval was quickly received, and the first Islander arrived to start these services on February 26 1968, and on March 1 the airline commenced full scheduled services between Alderney, Guernsey and Jersey. For the summer season, the airline flew three Islanders on these services, and for the following summer Aurigny introduced a walk-on no reservations system on April 1 1969. This proved to be an immediate success, and by the end of August 1969 Aurigny had carried one hundred thousand passengers on its services. This compared with twenty five thousand passengers carried during 1968, and traffic had increased to such an extent by 1969 that Aurigny was breaking records almost every single day. For instance, at the height of the 1969 season, Aurigny was offering ninety eight daily flights on its inter-island schedules, and on August 8 1969 the company carried a record total of six hundred and thirty two passengers on its services. On January 1 1970 Aurigny took over the former British United Island Airways routes from Alderney to Southampton; from Jersey to Dinard; and from Jersey to St.Brieuc. For the year up to the end of February 1970, Aurigny carried a total of 150,000 passengers and 300 tons of freight, and during the following summer up to one hundred daily inter-island schedules alone were operated. On November 16 1970, Aurigny opened a service between Alderney and Cherbourg using Islander G-AWBZ for the first service, and later during 1971 a service from Cherbourg to Southampton was also started.

In March 1971, Aurigny announced that it had ordered three Britten-Norman Trislanders for introduction into service later in the year. The airline's first Trislander arrived on June 29 1971, and on October 1 the aircraft entered scheduled service operating on the inter-island service between Jersey and Guernsey. On November 1, the Trislander was placed into service on the Guernsey - Alderney - Southampton service, and now the airline has six of these aircraft in service. Aurigny's average stage length on its inter-Channel Island sectors is a mere 26½ miles, and the Islander has proved itself to be ideal for these services. Aurigny undertakes little charter work, but the Islanders are occasionally used for air ambulance flights from the Channel Islands to the mainland.

Fleet:

BN.2 Islander	G-AVCN	c/n 003	dd. 8.67,	sold 4.75,	converted to BN.2A Islander.
	G-AWBY	c/n 017	dd. 3.68,	sold 1.72,	converted to BN.2A Islander.
	G-AWBZ	c/n 018	dd. 3.68,	sold 1.72,	converted to BN.2A Islander.
BN.2A Islander	G-AWNR	c/n 030	dd.10.68,	sold 12.71.	
	G-AWVY	c/n 048	dd. 3.69,	sold 1.72.	
	G-AXVR	c/n 139	dd. 2.70,	sold 1.73.	
	G-AXWP	c/n 147	dd. 2.70.		
	G-AXWR	c/n 149	dd. 2.70.		
	G-AYCV	c/n 170	lsd. 5.73,	ret. 9.73,	leased from Island Air Charter.
	G-BAVT	c/n 180	lsd. 5.73,	ret. 5.73,	leased from Intra Airways.
BN.2A Mk.111 Trislander	G-AYWI	c/n 262	dd. 6.71.		
	G-AZJA	c/n 305	dd.12.71.		
	G-AZLJ	c/n 319	dd. 1.72.		
	G-BBYO	c/n 362	dd. 3.74.		
	G-BCNO	c/n 1005	dd. 3.75.		
	G-BCXV	c/n 1007	dd. 5.75.		

AVIATION BEAUPORT

Aviation Beauport Ltd. Base: Jersey Airport.

Aviation Beauport was formed at Jersey Airport in June 1969 as an air taxi and executive charter operator. Operating a Cherokee Arrow, the airline acquired a new Piper Aztec D which joined the fleet on July 2-. Both aircraft were then used on many inter-island air taxi flights, and also on charters to towns in France and southern England. During 1969, Aviation Beauport was appointed a Piper dealer by CSE Aircraft Services, and the Cherokee Arrow was sold during 1970.

On August 6 1970,a Navajo was delivered to the company, and this aircraft was operated on behalf of Bayswater Services. This aircraft, the Aztec and a Baron have all flown numerous air taxi flights to the British Isles, the Aztec being the most common visitor to many airports including Birmingham, Gatwick, Hurn, Lulsgate, Southampton and Staverton. The company has also assisted many other charter and private companies in the operation of business aircraft.

Fleet:

Piper PA28R Cherokee Arrow 180	G-AWFC	c/n 28R-30670	dd. 6.69,	sold 9.69.
Piper PA27 Aztec 250D	G-AXKD	c/n 27-4293	dd. 7.69.	
Piper PA31 Navajo 300	G-AYFZ	c/n 31-679	dd. 8.70,	, operated for Bayswater Services.
Beech B55A Baron	G-AXXR	c/n TC-1347	dd. 73.	

BARDOCK

Southern Counties Aerial Contracts Ltd.
Bardock Aviation Ltd. Bases: Staverton and later Swansea Airports.

In June 1961, a company named Southern Counties Aerial Contracts started operations from Staverton Airport. Initially a Cessna 180 and a Tiger Moth were used for various duties, mainly crop spraying and parachuting, but in February 1962 a Dragon Rapide arrived, and although this was also used as a parachuting platform, it was also used for ad-hoc passenger and freight charters. In March 1963, Southern Counties changed its name to Bardock Aviation, which had been named after its founders - Mr.H.G.Barwell and Sir Bernard Docker. Bardock continued ad-hoc passenger and freight charters from Staverton, and later in 1963 applied to the A.T.L.B. for permission to operate a network of scheduled passenger services. These services were: Staverton - Swansea - Bristol - Land's End - Scilly Isles; Swansea - Bristol - Southampton; andStaverton - Bristol - Land's End - Scilly Isles. Not unnaturally, Bardock's proposed services were objected to by Mayflower Air Services, and in November 1963 the applications were turned down by the A.T.L.B. However, Bardock immediately appealed against the Board's decision, and in the hope of a successful appeal, the airline purchased a Dove and a Dakota for use on these services. The Dove arrived at Staverton during January 1964 and was used on several charter flights from Staverton, but the Dakota remained at Liverpool and was never delivered to the airline. However, once again the A.T.L.B. turned down Bardock's applications, and so in May 1964 the Dove was sold, and although charter flying continued for a short time afterwards with the Rapide and Cessna 180, all flying had ceased by the end of 1964. Before the cessation of operations, Bardock did fly several charters from Swansea and other airports in the South West of England and the South Wales area.

Fleet:

Cessna 180D	G-ARGC	c/n 50931	dd. 6.61,	ISOC.
DH82A Tiger Moth	G-ANEW	c/n 86458	dd. 6.61,	ISOC.
DH89A Dragon Rapide	G-AKIF	c/n 6838	dd. 2.62,	ISOC.
DH104 Dove 1	G-AIWF	c/n 04023	dd. 1.64,	sold 5.64.
Douglas C47B Dakota	G-AMNL	c/n 33392	never delivered.	

BARON

Baron Air Charter Ltd. Base: Southend Airport.

Baron Air Charter was formed as a subsidiary of General Engineering of Ilford, Essex. Initially the engineering firm operated a standard Beech Baron from June 1969 onwards, and this was used to carry the company's executives from Southend to Bristol and Derby. However, in order to fully utilise the aircraft, Baron Air Charter was formed in April 1970. The company's first commercial flight took place in May, and late in November 1970, Baron Air Charter took delivery of a new Baron 58. This aircraft is operated under the Air Operator's Certificate of Eagle Flying Services and is operated almost exclusively on executive charters.

Fleet:

Beech B.55 Baron	G-ASOH	c/n TC-656	dd. 6.69,	sold 11.70.
Beech B.58 Baron	G-AYGZ	c/n TE-112	dd.11.70.	

BAYLEE

Baylee Air Charter Ltd.
Baylee Air Holdings Ltd.

Base: Jersey Airport, with outstations at Guernsey and Exeter Airports.

Baylee Air Charter was formed early in 1971 by Captain Tony Baylee, and started air taxi and executive charter operations from Jersey in April with a second had Piper Aztec. As business built up, two further Aztecs were purchased during 1972, and with more and more flights being undertaken from Guernsey, one of the Aztecs was based there from 1973 onwards. Another of the Aztecs was based at Exeter from January 1973 onwards - the first aircraft based there was G-AYTC, but since then G-ATJZ has also spent much of its time at Exeter. Three of the aircraft have been named in a similar manner to those aircraft once operated by Jersey Airways before the last war, and in Baylee's smart blue and white colour scheme the aircraft are regular visitors to many English, French and Irish airports. In November 1974, Baylee Air Charter and Air London agreed a joint plan to operate emergency medical flights between the Channel Islands and the mainland - the main requirement being the carriage of blood plasma in 'life or death' cases. On November 19 1975, Baylee demonstrated its air ambulance Aztec to ambulance officials from Hampshire and Wiltshire at Hurn Airport. This service is now offered from Bournemouth, particularly to enable injured holidaymakers to the Bournemouth area to return to their homes by the quickest and easiest way. By the beginning of 1976 Baylee hoped to have three of their aircraft available for air ambulance operations anywhere within the British Isles.

During 1975, the company's aircraft have been particularly busy, with the Exeter based aircraft flying a regular air taxi service to White Waltham. In addition, one of the Jersey based aircraft has made numerous runs from Cork and Shannon to towns in north-western France including Quimper and St.Brieuc. On these flights to towns in north-western France the Aztecs carry fish and several flights have also been made from Haverfordwest to airports in Brittany carrying similar types of cargo. From Jersey and Guernsey, the aircraft make regular trips to Exeter, Gatwick, Hurn, Lulsgate, Plymouth and Rhoose. During the 1975 Admiral's Cup race in August, one of the Aztecs, operating from Plymouth, followed the course of this race taking photographs and checking the progress of the yachts.

Fleet:

Piper PA27 Aztec 250C	G-ATJZ	c/n 27-2967	dd. 4.72,	'St.Brelade's Bay'.
	G-AYTC	c/n 27-2725	dd. 4.71,	'Bonne Nuit Bay'.
	G-BADD	c/n 27-3122	dd.11.72,	reregistered G-BALU.
	G-BBAV	c/n 27-3173	dd. 7.73,	'St.Owen's Bay'.

BCI AIRWAYS

BCI Airways Ltd. Base: Barton Aerodrome.

BCI Airways was formed early in 1948 as the air charter division of British Cellulose Industries (Manchester) Ltd. Operations started in March with two Austers, a Dragon Rapide and a Miles Gemini. Based at Barton Aerodrome, BCI Airways operated passenger and freight ad-hoc charters in the north of England, and also undertook pleasure flying at Manchester's Ringway Airport during the summer of 1948. Pleasure flying was also undertaken at many air displays in the north of England, and one of the Austers (G-AIZX) was based at Southport, and during the summer operated pleasure flights from the beach. For the winter, BCI Airways suspended its pleasure flying activities from Southport, and although a small amount of air charter work was undertaken, all flying had ceased by the end of January 1949.

Fleet:

Auster J/1 Autocrat	G-AIPW	c/n 2204	dd. .48,	ISOC.
	G-AIZX	c/n 2231	dd. 3.48,	ISOC.
DH89A Dragon Rapide	G-AFMF	c/n 6432	dd. 3.48,	ISOC.
Miles M.65 Gemini 1A	G-AJEX	c/n 6324	dd. .48,	ISOC.

BEES FLIGHT

Bees Flight Ltd. Base: Sandown Airport, I.O.W.

Bees Flight was registered as an airline company on February 8 1947 by Mr.J.P.Stephenson and Mary Williams. The company opened an office at Carisbroke on the Isle of Wight, and operations started from Sandown later in the year with one Miles Gemini. During the summer Bees Flight used Miles Geminis mainly on pleasure flights from the airfield at Sandown, and at many displays the Geminis were used for joy riding.

By 1950 Bees Flight was flying a fleet consisting of one Dragon Rapide and three Miles Geminis on its pleasure flying duties. In addition the aircraft were often used by local aero club members, and many trips were made across the Channel to airfields in Belgium and France. Other charter flights were made to many airports in the British Isles, and the Geminis and Rapide were often used for air ambulance flights. In May 28 1950 Bees Flight's Gemini (G-AJFD) took part in the annual round the Isle of Wight air race, but its success, though, was not outstanding - it was placed ninth, and there were only nine entrants.

Bees Flight's operations, therefore, were a mixture of club and charter flying, and although the Geminis and Rapide were later sold, Bees Flight continued charter flying in the 1960s with two Austers. The company was also responsible for the operation of Sandown Airport, and when Sandown Airport closed in the late 1960s, Bees Flight, too ceased operations and the aircraft were sold.

Fleet:

Miles M.65 Gemini 1A	G-AJOL	c/n	6321	dd.10.47, written off Cowfold, Sussex 29.8.49.
	G-AJFD	c/n	6325	dd. 9.48, sold .50.
	G-AKHC	c/n	6490	dd. 5.50, sold 4.54.
	G-AKFX	c/n	6502	dd. 6.50, sold .55.
DH89A Dragon Rapide	G-AKMH	c/n	6704	dd. .50, sold .55.
Auster J/5B Autocar	G-AMPW	c/n	2961	dd. 5.52, sold 3.63.
Auster J/1N Alpha	G-APJZ	c/n	3382	dd. 1.63, ISOC.

BIDGOOD & CATTON

Bidgood and Catton Taxi Planes Ltd. Base: Sherburn-in-Elmet Airfield.

Bidgood and Catton Taxi Planes was founded on September 13 1948 by Messrs.J.C.Bidgood and D.Catton. Operations began from the former Royal Air Force airfield at Sherburn-in-Elmet with two Auster Autocrats and a Miles Messenger used on air taxi flights in the north of England. These three aircraft had all been owned privately by the firm's two founders before the formation of this charter company. Apart from the occasional air taxi flights, the aircraft were used for pleasure flying at local air displays and for flying training by the aero club at Sherburn-in-Elmet. In addition, the aircraft were still used by Messrs. Bidgood and Catton for their own private use.

By 1951 only the Miles Messenger remained in service, and over the next two years the operations of this company were run down, although the Messenger was not sold until 1958.

Fleet:

Auster J/1 Autocrat	G-AGVI	c/n	1860	dd. 9.48, sold 8.50.
	G-AHHM	c/n	2016	dd. 9.48, sold 10.49.
Miles M.38 Messenger 2A	G-AKIO	c/n	6729	dd. 9.48, ISOC.

BIRKETT

Birkett Air Service Ltd. Bases: Heston Aerodrome, and later Croydon Airport.

Birkett Air Service was formed by Mr.George Birkett as an air charter operator during 1932. From Heston Aerodrome, Birkett flew a fleet of Leopard Moths on various ad-hoc passenger and freight charters, including several on behalf of the press. The company had asked the Miles Aircraft Company to build for them a five seat aircraft capable of cruising at 140 mph, and this aircraft arrived at Heston late in May 1935 as the Miles Merlin. This aircraft entered service with Birkett, and was used on numerous long distance charters to Europe, the Middle East and North Africa. The company was also responsible for the operation of No.29 Elementary and Reserve Flying Training School at Luton, but on the outbreak of war in 1939 all operations were suspended.

However, in May 1947 services were resumed from Croydon Airport with two Dragon Rapides and one Leopard Moth. These aircraft were used on numerous ad-hoc passenger and freight charters, and the aircraft were also used for services to racemeetings. Close association was maintained with Olley Air Service who were responsible for the hangarage and servicing of Birkett's aircraft. On July 1 1950 Island Air Services opened a scheduled service between Shoreham and Deauville under a B.E.A.C. Associate agreement, and this service was flown jointly by Island Air Services and Birkett. However, this service proved to be uneconomic and was dropped in 1951, but further work came in 1951 in the form of the pleasure flying contract at Heathrow Airport on which Birkett's Rapides flew throughout the summer. In 1952, pleasure flying was carried out from Northolt with the Rapides, and also continued from Croydon. Early in 1953, Birkett was one of the first airlines to apply for permission to operate scheduled services, applying for routes from Croydon to Southampton and Alderney, and for a service from Southampton to Alderney specifying Rapides as the type of aircraft to be used. However, the airline started to run down its commercial flying operations at this time, and on July 31 1953 all flying by Birkett at Croydon Airport came to an end. Birkett still held several Royal Air Force training contracts, and these were either completed or passed to other operators.

Fleet:

Miles M.4 Merlin	G-ADFE	c/n	151	dd. 5.35
DH90 Dragonfly	G-AEDV	c/n	7524	bt.12.39, sold 3.40.
DH85 Leopard Moth	G-ACSH	c/n	7067	bt. 6.37, sold 3.40.
	G-ADAR	c/n	7091	dd. 2.35, written off Heston 10.7.35.
	G-ADBH	c/n	7030	bt. 4.36, sold 1.39.
	G-ADCO	c/n	7093	bt. 2.36, sold 3.40.
	G-AIYS	c/n	7089	dd. 3.47, sold 6.49.
Airspeed AS65 Consul	G-AJLK	c/n	5129	dd. .47, sold 6.50.
DH89A Dragon Rapide	G-AJBJ	c/n	6765	dd. 1.47, ISOC.
	G-AJDN	c/n	6860	dd. 4.47, ISOC.
	G-AKTZ	c/n	6482	dd. 5.48, ISOC.
Percival Proctor 4	G-ALEO	c/n	H.672	dd. 5.48, ISOC.

BLACK ISLE

Black Isle Air Services Ltd.
Capital Aervices (Aero) Ltd.

Bases: Inverness Airport (Black Isle), Glasgow and Edinburgh Airports (Capital).

In November 1961, a new company - Black Isle Air Services - applied to the A.T.L.B. for permission to operate a number of scheduled services from Inverness. The company was an associate company of Capital Services (Aero) Ltd., of Turnhouse and Renfrew who intended to supply two Piper Aztecs to Black Isle for use on their proposed scheduled routes. Capital Services had been formed some time before by Robert Whiteford, James Whiteford, James B.Whiteford and Norman Smith. This company operated from both Glasgow and Edinburgh and also operated several light aircraft on behalf of the Edinburgh Flying Club. A Piper Apache arrived at Renfrew Airport on June 12 1961 on delivery to Capital Services (Aero) Ltd. This aircraft was used by Capital for flying training from Glasgow and for air taxi work - during June and July the Apache visited several other airfields including Barra, Belfast, Edinburgh, Gatwick, Oxford, Perth, Prestwick and Rochester. Black Isle's applications were for services from Inverness to Glasgow, Edinburgh and Prestwick with up to six services a week in each direction. Also to be flown was a four times weekly service from Inverness via Stornoway to Benbecula; a twelve times weekly service from Inverness via Dornoch and Kirkwall to Sumburgh was also applied for. Black Isle stated that although a five seater Aztec would be used initially, as traffic built up, this aircraft would be replaced by a Dragon Rapide. However, the company's applications were subjected to very strong objections from both B.E.A. and the British Transport Commission, and in February 1962 Black Isle withdrew its applications and no flying ever took place. Capital Services was then taken over by Loganair on February 1 1962, and its chief pilot - Captain Duncan McIntosh - joined Loganair to set up their operations, which were to be run in a similar way to those proposed by Black Isle.

Fleet:

Piper PA23 Apache 160	G-ARBN	c/n	23-1385	dd. 6.61,	ISOT.

BLUE-AIR

Blue-Air Flying Services Ltd. Base: Blackbushe Airport.

Blue-Air Flying Services started air taxi and executive charter operations from Blackbushe Airport early in 1975 with one Cessna 310. This aircraft is operated by Brooklands Aviation on behalf of Blue-Air, and from Blackbushe this aircraft flies many charters to Continental destinations. In addition, Blue-Air Flying Services have also used the British Car Auctions Piper Aztec (G-BAVY) on some of their services, and the operations of Blue-Air are very closely associated with those of Brooklands Aviation. During 1975 the Cessna 310 regularly flew to destinations in France, Belgium and Holland, with the occasional trip being made to Scotland and Scandinavia. In addition, several charter flights were operated between Gatwick Airport and Gerona, in Spain.

Fleet:

Cessna 310Q	G-BBIM	c/n	0917	dd. .75.

BLUE-LINE

Blue-Line Airways Ltd. Base: Nottingham (Tollerton) Airport.

Blue-Line Airways was founded by a Nottingham businessman, Leslie M.Cooper, in the summer of 1946. Ad-hoc passenger and freight charters were undertaken from Nottingham Airport with two Proctors, and in May 1947 these two aircraft were joined by an Avro Anson. This aircraft flew many services to the Channel Islands from Nottingham during the summer of 1947 carrying holidaymakers. In addition to operating these tourist services to Jersey and Guernsey, Blue-Line Airways was also responsible for their organisation - the company also had opened travel agencies in Nottingham. Apart from their own Anson, Blue-Line Airways also used Air Transport Association's Ansons for these services, and the flights to the Channel Islands continued throughout the summer of 1947. For the 1948 season, Blue-Line Airways had five Ansons in service flying holidaymakers from Nottingham to Jersey and Guernsey, and the airline no longer used Air Transport Association's aircraft.

This expansion in 1948 caused Blue-Line Airways to be registered as an airline company on June 30 1948 by Leslie M.Cooper and Walter A.Andrews. This new company was set up to bring together the airline and travel agency parts of the Blue-Line empire. The airline continued operating its services to the Channel Islands throughout the summer, and between September 27 and November 21 Blue-Line operated a series of charter flights from Croydon to Le Bourget. These flights carried passengers from Croydon to Le Bourget, where they then boarded a Mercury Aviation aircraft in order to complete their journey to South Africa. These services were arranged by Ackroyds Air Travel, who were later fined for operating these services, and Blue-Line Airways thus lost this valuable source of income during the winter months. During the winter only the occasional ad-hoc passenger or freight charter was flown, but in January 1949 Blue-Line announced that it had bought a Dakota from B.O.A.C. which it intended to use primarily for freight charter work.

The Dakota entered service with Blue-Line at the end of February, but there was insufficient work in the Nottingham area to base the aircraft continually at Tollerton. Consequently, the Dakota moved south to the London area and then operated a variety of charter flights from Bovingdon and Blackbushe airports. These flights included the occasional service from Lille to Bovingdon carrying textiles on behalf of Lep Air Services during April and May. During June the Dakota transported soft fruit from Carmoli in Spain to Bovingdon, pigeons from Bovingdon to Brussels, and at the ned of the month Blue-Line operated a passenger charter flight from Kristiansand to Thornaby. On this flight, the Dakota carried twenty members of a ship's crew to Thornaby after clearing customs at Ringway en-route. By June the Dakota also occasionally appeared on the seasonal tourist run from Nottingham to Jersey - the Ansons were still used on this service, and during the summer, tourist services were also flown to the Isle of Man from Nottingham. One of Blue-Line's longest distance charters took place in the second week of July, when the Dakota was chartered by H.Clarkson (Air and Shipping Services, Ltd.) to carry twenty one passengers from Blackbushe to Nairobi. Shortly after its return from Nairobi, the Dakota was once again off on another long distance charter on behalf of H.Clarkson. On July 20 G-AGNG left London bound for Damascus, Kirkuk and Bahrein carrying cargo on behalf of the Iraq Petroleum Company.

This first flight on behalf of the Iraq Petroleum Company was followed by two further flights to the Persian Gulf and Damascus from London - one flight took place on July 28, and the second on August 3. At the height of the summer season, therefore, the Dakota was rarely used on the tourist services to Jersey, while the Ansons would spend most of their time plying between Tollerton and Jersey carrying holidaymakers. However, the summer had not proved to be profitable for Blue-Line, and on August 15 1949 Blue-Line announced that it was unable to continue flying owing to increasing losses on its services. The company, therefore, went into voluntary liquidation - twenty tourists were left stranded in Jersey, but were fortunately brought back by other charter companies. However, over two hundred passengers who had booked seats with Blue-Line for the remainder of the season, lost both their holidays and their money. Most of Blue-Line's aircraft were bought out by Eagle Aviation, who were, later, to operate scheduled services from Tollerton to Jersey.

Fleet:

Percival Proctor 5	G-AHWT	c/n	Ae.75	dd. 7.46,	ISOC.
	G-AIER	c/n	Ae.99	dd.10.46,	ISOC.
Auster J/1 Autocrat	G-AIGM	c/n	2177	dd. 2.47,	ISOC.
Miles M.65 Gemini 1A	G-AJTL	c/n	6461	dd. 9.47,	ISOC.
Avro 652A Anson 1	G-AJBA	ex.	AX409	dd. 5.48,	ISOC.
	G-AJFX	ex.	NK826	rd. 2.47,	ISOC.
	G-AKFK	ex.	NK770	ca. 3.48,	ISOC.
	G-AKFL	ex.	NK674	ca. 6.48,	ISOC.
	G-AKFM	ex.	MG495	ca. 4.48,	sold 6.49.
Douglas C47B Dakota	G-AGNG	c/n	26997	dd. 1.49,	ISOC.

BON-AIR

Bon-Air Flight Ltd. — Base: Aberdeen (Dyce) Airport.
Grampian Air Taxis Ltd.

Bon-Air Flight was formed at Aberdeen Airport in the summer of 1972, and took delivery of a Piper Aztec early in September for use on the North Sea oil rig support work. However, the first aircraft was badly damaged while landing at Nigg Bay after only a few days of operations, and a replacement Aztec was quickly found. The company also traded as Grampian Air Taxis, and it was under this title that the company operated a third Aztec (G-BAED), initially flown on lease from Southend Air Taxis. From Aberdeen, the airline flew many charters to Sumburgh, and across the North Sea to Norway and Denmark, but in August 1973 the airline was bought out by Airgo, another Scottish air taxi operator.

Fleet:

Piper PA27 Aztec 250B	G-ASER	c/n 27-2283	dd. 9.72, written off Nigg Bay 14.9.72.
Piper PA27 Aztec 250C	G-ASRE	c/n 27-2520	dd. 9.72, sold 4.73.
	G-BAED	c/n 27-3864	dd.11.72, ISOT.

BOND AIR SERVICES

Bond Air Services Ltd. — Bases: Sywell Aerodrome, Southend and Gatwick Airports.

Bond Air Services was formed at Gatwick Airport by Mr.D.E.Bond in the summer of 1946 as a general air charter operator. Initially both passenger and freight charters were flown with a Percival Proctor and an Auster - the passenger charter work included pleasure flying during the summer months. In March 1947, though, Bond Air Services introduced a Halifax into service and this aircraft was used during the spring to transport soft fruit, including apricots, greengages and peaches from towns in Italy, Spain and southern France to Bovingdon. By May, two of these aircraft were in service, but one of these aircraft was lost in an accident at Barcelona at the end of the month while collecting another cargo of fruit. The Proctor was also lost in an accident during May, but a DH86B was aquired in July and this aircraft was used for many passenger charters from Gatwick Airport.

On October 22 1947 Bond Air Services was taken over by Messrs.R.D.P.Gilmore and Robert A.Treen - both men became directors of the company, while Captain C.Treen became Bond Air Services' chief pilot. This take-over was followed by an expansion of the company's activities with the acquisition of a further Halifax and another DH86B. Two Halifaxes were now in service and these aircraft flew numerous ad-hoc freight charters during the latter part of 1947 including the carriage of ships' boiler pipes to Gibraltar, eggs from London to Paris and cheese and cloth to Casablanca. In addition, seven flights took place between Caen and Gatwick carrying Camembert cheese, and five flights between Valencia and Gatwick carrying manderines. During the spring of 1948 the Halifaxes imported a large amount of soft fruit once again from towns abroad, and this activity was followed in August by Bond's biggest ever operation.

On August 4 1948 one of Bond's Halifaxes started flying sorties on the Berlin Airlift from Wunsdorf. It was at this time that Bond Air Services became closely associated with Laker's Southend based Aviation Traders. Aviation Traders had purchased the entire fleet of B.O.A.C. Haltons, and many of these aircraft were then converted for use on the airlift and leased to Bond Air Services. Bond's one Halifax based at Wunsdorf was, soon joined by many other aircraft and at the height of the airlift Bond Air Services had twelve aircraft in service flying along the corridors into Berlin carrying essential supplies. On November 14 Bond moved their fleet of Halifax freighters from Wunsdorf to Hamburg, and operations continued from there until August 15 1949 when Bond was withdrawn from the airlift after flying a total of two thousand five hundred and seventy seven sorties in almost six and a half thousand flying hours. Bond had also purchased a Viking in December 1948 for use on the airlift, but with the withdrawal of twin engined aircraft from the airlift, the plans to use this aircraft were dropped and the Viking never entered service. While Bond had been engaged in the airlift in Germany, little other charter work was undertaken although one Halifax was released for the milk run from Northern Ireland to Northern England. On this airlift, Bond Air Services operated one Halifax between August 31 1948 and October 31 1948 carrying up to seven hundred gallons of milk on each trip from Nutts Corner, Belfast to Liverpool and Blackpool.

Both airlifts now over, the fleet of Halifaxes and Haltons returned to Southend, their task completed. Bond Air Services were now based at Southend, although a base was also opened at Rearsby from where Bond flew passenger and freight charters with the DH86s. The depression in charter work after the airlift caused Bond to drastically reduce its fleet. In Germany, Bond had flown its aircraft at an all up weight of fifty nine thousand pounds, but on its return to England Bond was only permitted to operate its aircraft at a weight of fifty seven thousand pounds - thus reducing the payload of the Halifaxes by almost one ton. Immediately after the end of the airlift, Bond Air Services was able to offer five Haltons and Halifaxes for ad-hoc freight charter work from Southend. However, with little work available this number was reduced to two aircraft by the beginning of 1950, and although the aircraft did find work transporting textiles from Lille to Southend and importing fruit from abroad, Bond ceased all Halifax flying in May 1950. The airline then turned its attentions to passenger and freight charter work with two DH86s, and these aircraft were mainly flown from Gatwick and Rearsby.

Bond Air Services also operated the occasional charter flight from Skegness Airport - one of Bond's owners, Mr.Robert Treen, and Mr.F.A.Laker had founded Skegness (Airport) Ltd., on September 9 1949. This company was responsible for the operation of Skegness Airport, and charter flights were often flown by Bond Air Services on behalf of Skegness Airport. During 1950 an Air Display was held at Skegness, and Bond's DH86s were used for pleasure flying at this event. These aircraft were also often used for charter flights to the Continent, and for services to race meetings around the British Isles.

Bond was, by 1951, an air taxi operator, and the two DH86s proved to be too large for the most of their operations. These aircraft were, therefore, sold and during 1951 Bond operated air taxi flights with a Dragon Rapide and a Fairchild Argus, but all operations had ceased by the end of 1951.

Fleet:

Auster J/1 Autocrat	G-AGYG	c/n	1988	dd. 6.46, sold 6.49.
Percival Proctor 1	G-AHMS	c/n	K.248	dd. 7.46, written off Tonnere, France 14.5.47.
Handley Page 61 Halifax 8	G-AIOH	c/n	1324	dd. 5.47, written off Barcelona 30.5.47.
	G-AIOI	c/n	1327	dd. 3.47, written off Tegel, Berlin 15.2.49.
	G-AIWW	c/n	1319	dd. 3.48, wfu. 5.50.
	G-AIZO	c/n	1366	dd.10.47, written off Studham 23.5.48.
Handley Page 61 Halifax 9	G-ALON	c/n	1451	dd. 6.49, wfu. 6.50.
	G-ALOS	c/n	1589	dd. 6.49, wfu. 7.49.
Handley Page 70 Halton	G-AHDN	c/n	1318	lsd. .48, ret. 3.49, leased from Aviation Traders.
	G-AHDO	c/n	1320	lsd. 3.49, ret. 7.49, leased from Aviation Traders.
	G-AHDP	c/n	1341	lsd. 7.48, written off Schleswigland 9.4.49, leased from Aviation Traders.
	G-AHDS	c/n	1350	lsd. 7.48, ret. 7.49, leased from Aviation Traders.
	G-AHDT	c/n	1370	lsd. 7.48, wfu. 12.48, leased from Aviation Traders.
	G-AHDU	c/n	1372	lsd. .48, wfu. 4.50, leased from Aviation Traders.

Handley Page 70 Halton	G-AHDW	c/n	1377	lsd. .48, wfu. 12.48, leased from Aviation Traders.
	G-AHDX	c/n	1378	lsd. .48, ret. .49, leased from Aviation Traders.
DH86B	G-ADUH	c/n	2336	dd. 7.47, sold 2.51.
	G-ADVJ	c/n	2338	dd.10.47, sold 2.51, 'County of Sussex'.
Vickers 614 Viking	G-AHPE	c/n	137	dd.12.48, sold 1.49, never entered service.
DH89A Dragon Rapide	G-AKOG	c/n	6878	dd.11.50, ISOC.
Fairchild Argus 3	G-AJPE	c/n	344	dd. 5.51, ISOC.

BOSTON

Boston Air Transport Ltd. Base: Boston Aerodrome.
Boston Aero Club Ltd.

Boston Air Transprt was a small air charter company founded on December 17 1948 by a group of businessmen. This group consisted of Messrs.Charles D.Allison of Nottingham, Ernest W.Pearson of Boston and Edward A.Moffatt also of Boston. Mr.Moffatt was also a director of the Boston Haulage Company, and Boston Air Transport was really an extension of this road haulage company. During 1949 the company operated a fleet consisting of two Miles Messengers and a variety of Auster types, mainly on pleasure flying and flying instruction from Boston Aerodrome. The company was also responsible for the operation of the Boston Aero Club which used all Boston Air Transport's aircraft. By 1950 the membership of this club exceeded seventy people - the club's President was Mr.R.Hardy, while the Secretary was Mr.E.W.Parson, AFC. This fleet remained unchanged until December 1950 when one of the Austers was lost in an accident, followed by the sale of another aircraft to France in July 1951. In addition to pleasure flying and flying tuition, Boston Air Transport also undertook light passenger and freight ad-hoc charter work. Boston had hoped to increase this type of work during 1950 with the acquisition of a Rapide and a Gemini, but this expansion never took place, and operations came to an end at the close of the 1954 summer season.

Fleet:

Auster J/1 Autocrat	G-AGYR	c/n	2008	rd.11.50, written off Kirkburton 11.12.50.
	G-AIPU	c/n	2202	rd. 6.49, sold 7.51.
Auster J/4	G-AIRD	c/n	2086	rd.10.49, ISOC.
Taylorcraft Plus D	G-AHCG	c/n	206	dd. .49, ISOC.
Miles M.38 Messenger 2A	G-AJDM	c/n	5347	rd. 8.49, ISOC.
	G-AKKK	c/n	6712	dd. 4.49, sold 8.54.

BOURNAIR

Bournair Ltd. Base: Bournemouth (Hurn) Airport.
Bournemouth Air Taxi Company Ltd.

The Bournemouth Air Taxi Company was formed in November 1961 with one Auster J/5 to operate air taxi services in the Bournemouth area. Early in 1962 a Cessna 172 was acquired, and this aircraft and the Auster flew joy rides from Bournemouth throughout the summer of 1962, and the Cessna also acted as crew ferry in April when it flew to Blackpool to collect a Dragon Rapide for the company. When the airline had been formed, it applied to the A.T.L.B. for permission to operate a scheduled passenger service between Bournemouth and Sandown. This was approved by the Board early in 1962, for operation over this route four times daily by Rapide from April onwards. In addition to this, the Rapide was used for a large amount of pleasure flying, but in May 1964 it was exchanged for a Dove. This rather more modern aircraft operated air taxi services from Bournemouth to many parts of the British Isles, and also to Northern France. Many trips were also made to the Channel Islands, but in July 1967 the aircraft was sold. By this time, the charter side of the business had been run down, and the company now exists as part of the Bournemouth Flying Club, and concentrates mainly on flying training, although a small amount of pleasure flying and charter work is still carried out with the club aircraft.

Fleet:

Auster J/5	G-ARLY	c/n	3271	dd.10.61, sold 11.65.
Auster J/1N Alpha	G-AGYT	c/n	1862	rd. 7.66, sold 11.66.
Cessna 172	G-ARID	c/n	48209	dd. .62, sold .
Cessna 185	G-ATAC	c/n	0807	dd. 1.65, sold 2.65.
Cessna 310	G-ATCR	c/n	35311	dd.10.64, sold .
DH89A Dragon Rapide	G-AKRS	c/n	6952	dd. 4.62, sold 5.64.
DH104 Dove 1B	G-AOVY	c/n	04009	dd. 5.64, sold 7.67.

BREVET

Brevet Flying Club Ltd. Bases: Hanworth Aerodrome and Croydon Airport.

The Brevet Flying Club was formed in 1943 and, based initially at Hanworth Aerodrome, this company undertook flying training with a variety of light training aircraft from 1946 onwards. Passenger charter work was also undertaken with a Percival Vega Gull from January 1946 onwards. However, this aircraft was damaged in an accident at Elmdon Airport on May 15 1946, and the aircraft's remains were sold to a company at Gravesend, where the aircraft was then rebuilt. The Vega Gull was replaced by an Airspeed Envoy which flew several passenger charters from both Hanworth and Croydon, and the company advertised that it could operate a charter service 'within a radius of five hundred miles of London'. The Envoy was sold at the end of the 1946 season, and operations continued over the next two years operating a variety of flying club single engined aircraft. In 1948 a Consul was purchased for use on charter flights from Croydon, and although this aircraft was advertised for sale during 1949, the aircraft remained in service until being scrapped early in 1951. The Brevet Flying Club also played an important part as one of the top aviation social clubs in London - its headquarters were in London's Mayfair, and here the flying club offered a fully licensed residential clubhouse, with excellent food. By 1950 its membership had exceeded ten thousand people, but air charter operations came to an end early in the following year. The social club, though, continued in existence and the club became an associate of the Brevet Club of New Zealand, and the Aircrew Club of Australia.

Fleet:

Percival Vega Gull	G-AEXV	c/n	K.57	dd. 1.46, sold after accident at Elmdon 15.5.46.
Airspeed AS6J Envoy Series 111	G-AHAC	c/n	79	rd. 1.46, sold 7.46.
Airspeed AS65 Consul	G-AIRP	c/n	4400	dd. .48, wfu. 4.51.

Britavia's Hermes were often used for inclusive tour flights from Blackbushe to Austria, France, Italy, Spain, Switzerland and Yugoslavia - here G-ALDU taxies in after completing another trip.

Michael Stroud

British Caledonian's one piston-engined aircraft is this very smart Piper Aztec, which is used for crew positioning, training and other flights under the R/T callsign BR 049.

Gordon Bain

One of eight Super 1-11's absorbed into the British Caledonian fleet from British United Airways, G-AWYS is shown here in flight with intermediate Caledonian//BUA markings.

British Aircraft Corporation

British Air Ferries' Carvair fleet is now dwindling as aircraft are gradually replaced by Heralds, but G-ASHZ was still in service at the end of 1975.

Tony Merton Jones

British Air Ferries only used two CL-44's for a short time due to operating problems with the type, and the aircraft are now back in service with Trans Meridian.

Ian MacFarlane

British Air Ferries' first Herald G-BCWE is seen here on July 31 1975 after operating an Ostend to Southend scheduled service.

Tony Merton Jones

G-AFYH British Air Transport Flamingo — A J Jackson

G-AKVW British Air Transport Anson — A J Jackson

G-AMLP British United Air Ferries Freighter — Ian MacFarlane

G-ASJE Caledonian//BUA BAC 1-11 srs.201 — Ian MacFarlane

N501TL Alidair Viscount 812 — P H Butler

EI-AOI British Air Ferries Viscount — Austin J Brown

BRISTOL AIR TAXIS

Bristol Air Taxis Ltd.
Lansbury Aviation Services Ltd.

Bases: Bristol (Lulsgate), Weston-Super-Mare and Staverton Airports.

Bristol Air Taxis was formed in April 1972 at Lulsgate Airport jointly by Lansbury Aviation Services and Lulsgate Air Services. Lansbury contributed its one Aztec to the new fleet, and Lulsgate one Aztec and a Twin Comanche. The latter two aircraft initially flew with both Lulsgate Air Services' titling and Bristol Air Taxis' titling, but soon all aircraft were flying in the standard Bristol Air Taxis livery. Gradually the older Aztecs were replaced by a fleet of new Aztec Es, and the company also has the use of Unigate's Aztec which is normally based at Bristol. An office was opened at Staverton during 1973 to supplement the company's already well established bases at Lulsgate and Weston-Super-Mare.

During 1974 regular flights were made from Staverton and Lulsgate to Eindhoven in Holland, and from Lulsgate, Cardiff and Staverton to Antwerp in Belgium. During 1975 the fleet of Aztecs flew numerous air taxi flights from Bristol, Cardiff, Staverton and Weston to destinations all over the British Isles and Europe. From Bristol and Cardiff many flights were made to Scottish airfields including Aberdeen, Edinburgh, Glasgow and Prestwick. From Cardiff and Hurn several flights were made to Castlebridge in Southern Ireland, and also from Cardiff and company flew an occasional charter to Perranporth and Jersey. The Aztecs were also regular visitors to Luton and Southend Airports, and many destinations in northern France were served during the year. International charters included flights to Dinard, Le Havre, Amsterdam, Rotterdam, Hanover, Hamburg, Basle and Geneva. Bristol Air Taxis also undertakes work on behalf of the British Aircraft Corporation and the Vickers Group, and these contracts took Bristol Air Taxis' aircraft to Barrow, Filton, Hawarden and Warton.

Bristol Air Taxis is, therefore, one of Britain's major air taxi operators now flying a fleet of modern twin engined aircraft on charters throughout the British Isles and Europe.

Fleet:

Piper PA30 Twin Comanche	G-ASYK	c/n 30-573	dd. 4.72, sold .73.	
Piper PA27 Aztec 250C	G-ATFF	c/n 27-2898	dd. 4.72, sold 10.73.	
	G-AVTS	c/n 27-3489	dd. 3.73, sold 5.74.	
Piper PA27 Aztec 250D	G-AYWY	c/n 27-4069	lsd. , ret. ,	operated on lease from Unigate.
	G-AWVW	c/n 27-4054	dd. 4.72, sold 9.74.	
Piper PA27 Aztec 250E	G-BBOM	c/n 27-5208	dd. 5.74	
	G-BBRA	c/n 27-5197	dd.11.73	
	G-BBYK	c/n 27-5257	dd. 3.74	
Piper PA31 Navajo 300	G-AWOW	c/n 31-229	dd.10.73, sold 4.74.	

BRITANNIA

Britannia Airways Ltd.
Euravia (London) Ltd.
Skyways Ltd.

Base: Luton Airport.

Britannia Airways is now one of Europe's largest charter airlines, flying holidaymakers from many parts of the British Isles to all parts of Europe, North Africa and the Canary Islands with a fleet of Boeing 737s. Britannia Airways' history can be traced back to December 1 1961 when three people - Mr.J.E.D.Williams, Captain J.C.Harrington and Captain 'Ted' Langton - founded a new British charter airline named Euravia (London) Ltd. Captain Langton was the owner of a travel company named Universal Sky Tours, which had been flying holidaymakers to many Mediterranean sunspots for many years. During 1961 the company had employed several British charter airlines to operate its inclusive tour charter flights, mainly from London's Gatwick Airport, but two of these companies (Falcon Airways and Air Safaris) had run into financial and licensing difficulties, and had been forced to cease operations. As a result of these airline failures, Captain Langton decided to form his own airline to carry Universal Sky Tours' clients on their holidays, and thus Euravia was born.

The name of Euravia had been chosen as this airline's name because of Britain's proposed entry into the Common Market at that time. During the winter of 1961 Euravia evaluated several different types of aircraft for its 1962 operations, and finally in the early spring of 1962 it was decided to purchase a fleet of Lockheed Constellations. The ex-El Al 049 Constellations were acquired from the Israeli airline, and the first aircraft arrived at Luton Airport for Euravia on April 12 1962. The company had chosen Luton Airport as its main operations base for several reasons, including the airport's proximity to the newly opened M1 Motorway, and also because of the excellent hangarage and other airport facilities available to Euravia at Luton. On April 17 a 'welcoming' ceremony was held at Luton Airport in Euravia's honour, and this event was attended by the Mayor of Luton and many other civic dignitaries. During the next fortnight the airline carried out many route proving and crew training flights, but on May 5 1962 one of the Constellations (G-ARVP) flew Euravia's first revenue earning flight when it carried a load of inclusive tour passengers from Manchester to Palma. For the remainder of the summer the fleet of aircraft was kept busy flying tourists from Birmingham, Luton and Manchester to many Mediterranean resorts including Gerona, Malaga, Palma, Perpignan, Rimini,Tarbes and Valencia. These services were flown on behalf of several tour operators - apart from Universal Sky Tours, these tour operators included Ellis Travel and Midland Air Tours.

Of the personalities involved with Euravia, Ted Langton remained as Managing Director of Universal Sky Tours, while Mr.J.E.D.Williams became Managing Director of Euravia, and Captain John Harrington became a co-director of the airline. Mr.Williams also owned and managed a firm of aviation consultants named J.E.D.Williams and Company (Aviation Consultants) Ltd. Captain Harrington had previously flown with Imperial Airways from 1931 and then with B.O.A.C. until retiring from the position of Deputy Operations Director with the Corporation in 1956. He had then joined the Bristol Aircraft Company as General Services Manager, and had left the company during 1961 in order to assist in the formation of Euravia. Captain D.H.Davison acted as Chief Pilot with Euravia, while M.S.Finnigan became the company's Chief engineer.

Euravia's first summer proved to be an extremely healthy one, and towards the end of the season it was realised that further aircraft would be needed to cope with the expected increase in passengers during the 1963 season. During 1962 Skyways puchased the three Constellation that they had operated on lease from B.O.A.C. since June 1959. However, Skyways had found that work for these aircraft was difficult to find - a B.O.A.C. freight contract from London to Singapore had ended early in 1962, their trooping contracts had all passed to other airlines with more modern equipment, and the aircraft were now amost solely used on ad-hoc passenger and freight charters and on Skyways' scheduled services from London to Cyprus, Malta and Tunis. These services used Stansted as their London terminal and were known as 'Crusader' services, while from Heathrow Skyways operated a handful of veteran Avro Yorks on ad-hoc freight charters, mainly carrying spare parts and engines around the world on behalf of B.O.A.C. and Pan American. Skyways was, therefore, on its last legs financially, and in July 1962 negotiations began between Euravia and Skyways about the possible take-over of Skyways by Euravia. And so it was that, on September 1 1962, Euravia took over the assets of Skyways - this take-over included Skyways' fleet of aircraft and services, but not the company's debts. The Yorks remained in Skyways' colours and continued to fly on their ad-hoc freight charters from Heathrow. However, the Constellations were painted up in Euravia's colour scheme, and during the 1963 season the fleet of Constellations totalled an impressive eight aircraft. These aircraft were used on Euravia's inclusive tour flights, and also on Skyways' Crusader services to Cyprus,Malta and Tunis, and at least one of Euravia's Constellations was painted up in Skyways' colours for use on these services. On May 19 1963 the Constellation G-ALAL operated an inclusive tour flight from Newcastle to Valencia, marking Euravia's first inclusive tour flight from this airport - services continued throughout the

1963 season from Newcastle, and have done so ever since. During 1963 Euravia also operated a fortnightly inclusive tour flight from Liverpool to Valencia, but the services were not always as punctual as they might have been, and at the end of the season Euravia discontinued its services from Liverpool.

1964 was another year of chance for the airline - the routes to Cyprus, Malta and Tunis, which had been flown on a scheduled basis, were withdrawn altogether in April, and this event marked the last operation of any form of scheduled service by the airline. In addition, Euravia started flying holidaymakers to new resorts including Ibiza, Klagenfurt and Ljubljana. The airline was now flying tourists regularly from Luton, Manchester and Newcastle, but the services from Birmingham had stopped. The occasional inclusive tour flight was also operated from Blackpool, and during 1964 the Constellations flew inclusive tour services from Cardiff Airport to Barcelona, Palma and Perpignan. The Constellations , though, were now becoming too small for many of the airline's routes, and a replacement was then sought. B.O.A.C. had retired their entire Britannia 102 fleet and placed all the aircraft in storage at Cambridge, and although several airlines showed interest in purchasing these aircraft, Euravia obtained several of these aircraft late in 1964. Shortly before the introduction of the first Britannia, Euravia changed its name to Britannia Airways on August 16 1964, and the introduction of the Britannia fleet was assisted by the take-over of the airline by the Thomson Organisation on April 25 1965. Prior to this, in November 1964, the first Britannia was delivered to Britannia Airways at Luton Airport, and this aircraft (G-ANBB) operated Britannia Airways' first Britannia flight on December 6 1964 when it flew a party of holidaymakers from Luton to Tenerife.

By the summer of 1965 a total of five Britannias were in service with Britannia Airways, and by August all the Constellations had been withdrawn. Universal Sky Tours was still the airline's major customer, and this association was to continue, as the Thomson Organisation had also taken over Universal Sky Tours. The Britannias soon proved to be extremely popular with the airline's passengers and crews - equipped with one hundred and twelve seats, these four engined turboprop aircraft, affectionately named 'Whispering Giants', chopped well over an hour off most of Britannia Airways' inclusive tour routes. Two of the Constellations were sold to Aviation Charter Enterprises in August 1965, while a third aircraft was leased to ACE, returning to Britannia Airways on September 14 1966 when ACE ceased operations. For the 1966 summer season Britannia Airways had eight Britannias in service, but on September 1 1966 this number was reduced to seven after one of the aircraft was lost, in Britannia Airways' only fatal accident at Ljubljana. Just after dusk on August 31 1966, Britannia G-ANBB (the airline's first Britannia to enter service) took off from Luton Airport with seven crew and one hundred and ten passengers on board for a normal inclusive tour flight to Ljubljana in Yugoslavia. The aircraft climbed to its cruising level of FL210, and the flight proceede normally - the Yugoslav FIR boundary was passed at 0029 hrs, and the aircraft then set course for the Dolsko beacon descending to 4,500 feet. After checking this beacon the captain reported that he had the runway lights in sight, and an approach was then made to Runway 31. The pilot then continued the approach normally until, at a point 2½ miles from touchdown, when he requested radar assistance. However, little over one mile later the aircraft disappeared from radar, and G-ANBB had crashed into a wood one mile from touchdown. Of the one hundred and seventeen people on board, only nineteen passengers and one crew member survived, and most of these were in the tail section of the aircraft. This accident, at the time, was the worst accident to involve a British transport aircraft.

The 1968 summer season proved to be a busy time for Britannia, and several other aircraft were leased to the company by other airlines, and services were flown on behalf of many tour operators including Caytours, Global, Horizon Holidays, Luxitours, Riviera Holidays and Sky Tours (the 'Universal' had now been dropped). Numerous destinations were now served in Bulgaria, Greece, Italy, North Africa, Spain, Turkey, and Yugoslavia. The Britannias were now flown in a one hundred and seventeen passenger seat configuration, and extra aircraft had to be leased from Laker Airways and B.K.S. Air Transport due to the late delivery of Britannia Airways' fleet of Boeing 737s. Britannia Airways became the second European operator of the Boeing 737 - the first European airline to fly the 737 was lufthansa, and thus Britannia Airways was the first European charter airline to operate the type. Britannia Airways' Boeing 737s are of the 200 series variant, and these aircraft were initially fitted with one hundred and seventeen passenger seats in a six abreast seating configuration, although the seating capacity was increases to one hundred and twenty four seats in 1969, and to one hundred and thirty seats by the 1971 summer season. The first aircraft (G-AVRL) arrived at Luton on July 8 1968, received its Certificate of Airworthiness two days later, and on July 19 G-AVRL operated a proving flight from Luton to Palma and return. On July 22 a press/publicity flight was operated from Luton to Dubrovnik, and after a nightstop, the 737 returned to Luton on the following day. The aircraft then entered service on many of Britannia's services from Luton, and on August 16 G-AVRM entered service when it flew a Sky Tours service from Luton to Venice and return. In May 1968 Britannia Airways started flying inclusive tour charter flights from Bristol Airport on behalf of the Global Travel Company.

By the 1969 summer season Britannia Airways had five Boeing 737s in service, and at the end of the season the first Britannias were withdrawn from service. Over the past four years these aircraft had operated several ad-hoc passenger and freight charters to many destinations and several flights were made to North America. In addition several affinity group charters were flown from Luton to the Far East, and some work was also undertaken on behalf of the Ministry of Defence transporting troops and military equipment around Europe, the Middle and Far East. During the spring of 1969 Britannia Airways was awarded a one and a half million pound contract by the Ministry of Defence for the carriage of troops and their families between Luton and several Royal Air Force bases in West Germany. Since 1969 Britannia Airways' 737s have flown numerous services each week from Luton to airfields including Dusseldorf, Gutersloh, Hanover and Wildenrath, and a once weekly flight is now operated to Berlin.

For the 1970 summer season Britannia Airways had a fleet of four Britannias and eight Boeing 737s in service, and these aircraft flew services from Luton to Alicante, Genoa, Gerona, Ibiza, Malaga, Malta, Munich, Naples, Palma and Rimini; from Birmingham to Alicante, Gerona, Ibiza, Mahon, Malta, Palma, Rimini, Tunis, and Venice; from Bristol and Cardiff to Alicante, Gerona, Ibiza and Palma; from Glasgow to Alicante, Barcelona, Gerona, Ibiza, Palma and Venice; from Manchester to Alicante, Gerona, Ibiza, and Palma; from Gatwick to Gerona and Palma; from Liverpool to Palma; and from Newcastle to Alicante, Barcelona, Gerona, Ibiza and Palma. These services were almost exclusively flown by the 737s, but the four Britannias that were retained, were also kept busy. In addition to flying Ministry of Defence flights and ad-hoc passenger and freight charters, the Britannias were also actively engaged in carrying tourists from this country to the Oberammergau festival in West Germany. The Britannias flew once weekly services to Munich for this festival from Birmingham, Edinburgh, Gatwick, Glasgow, Liverpool, Luton, Manchester, Newcastle and Teesside. In March 1970 Britaania Airways took delivery of its first convertible passenger/cargo Boeing 737, and the first freight service flown by this aircarft was operated on March 21 1970 when G-AXNA flew from Luton to Dublin and back. Subsequently, the 737s were often used for carrying racehorses around Europe, and from Luton the aircraft made many flights to Dublin, Le Bourget and Milan carrying racehorses including Niginsky and Brigadier Gerrard. In addition Britannia Airways were responsible for carrying the British Show Jumping Team to Kiev. The 737s were also often used for long range charter flights - a weekly service, usually flown by G-AWSY, was flown to Hong Kong, Colombo and Bangkok. Another affinity group charter trip regularly flown by Britannia Airways, was a service to Kuala Lumpur from Luton which operated via Zagreb, Damascus, Dubai, Karachi and Bangkok.

At the end of 1970 the last survivor of the once proud Britannia fleet was withdrawn when G-ANBL flew a Thomson Holidays' flight from Luton to Genoa and return on December 29 1970. This was the last flight by a Britannia Airways Britannia, and the airline was now an all-jet operator. 1971 saw futher expansion with the delivery of a Boeing 707 when Britannia Airways took over the lease of World Airways' 707 N375WA early in the year. This was followed by the start of inclusive tour charter operations from Castle Donington - on April 17 1971 G-AXNB flew a service from Castle Donington to Palma, and G-AVRM flew a service from Castle Donington to Ibiza - both flights operated on behalf of Horizon Midlands, and these flights have continued ever since Britannia Airways' first Boeing 707 (G-AYSI) flew from Oakland to Luton on delivery to Britannia on February 28 1971, and then then entered Britannia's newly constructed hanger at Luton Airport for repaint and conversion for trans-Atlantic charter operations. Although this aircraft had been primarily acquired for use on trans-Atlantic affinity group charters, the Thomson Holidays' brochures all proudly announced that they would be using both Britannia Airways' Boeing 707s and 737s for their services that summer. The 707 entered service in April and in addition to flying affinity group charters to North America, the aircraft also operated inclusive tour charter flights from Gatwick, Glasgow, Luton and Manchester. In addition, several flights were operated from other provincial airports, and the acquisition of this type of aircraft allowed Thomson Holidays to organise holidays in far-away places such as in the Caribbean. However, the main role of the 707 was the operation of trans-Atlantic charter flights and other long range charter work. The majority of these trans-Atlantic affinity group charter flights were operated from Luton to Los Angeles with a stop at Keflavik for refuelling on the outward flight - technical stops were also occasionally made at Prestwick and Stansted Airports. Split load charters were also operated from Manchester and Prestwick to North America, particularly during the summer months.

A second Boeing 707 was acquired on October 1 1971 on lease from Executive Jet Aviation, and during the summmer of 1972 the two 707s operated many trans-Atlantic charter flights from Luton, Manchester and Prestwick to Los Angeles, and also several inclusive tour charter flights on behalf of Thomson Holidays. The 707s were used on a fortnightly holiday charter flight from Gatwick to Montego Bay, with one refuelling stop en-route, for Thomson Holidays, and the aircraft also appeared at Heathrow on several occasions operating flights on behalf of other airlines. These flights included the operation of several cargo services between Heathrow and Nairobi on behalf of East African Airways' charter subsidiary, Simbair. Other cargo charters included several flights from the United Kingdom to Tokyo carrying racehorses, and the 707s visited many parts of the world on a variety of charters. On November 2 1972 Britannia Airways became the first British airline to operate inclusive tour charter flights to Moscow. On this date 707 G-AYSI left Luton for Moscow as BY663 with the first party of Thomson Holidays' clients to visit the Russian capital city. On the return flight, the aircraft flew back empty, which was fortunate because due to adverse weather conditions at Luton, the 707 was forced to divert to Birmingham. Although the Boeing 707s had been making money for Britannia Airways, new charter regulations for the operation of trans-Atlantic charter flights caused Britannia to cut short their leases on the two aircraft and return them to their owners. These new regulations, which were introduced early in 1973, would have required Britannia Airways to have set up its own offices in North America, and the relatively heavy capital outlay was not practical for Britannia's comparatively small trans-Atlantic charter operations. The 707s were, therefore, returned to their owners, and Britannia Airways became an all Boeing 737 operator for the 1973 season.

During 1973, with eleven 737s in service, Britannia Airways flew inclusive tour charters from all normal British airports, although Blackpool was no longer served with its once weekly flight to Palma which had been flown by Britannia Airways for several years. The airline was now increasingly using Gatwick for many of its inclusive tour services, and the airports of Birmingham, Castle Donington and Manchester had also become popular departure points for Thomson's holidaymakers. In December 1973 Boeing 737 G-AVRL left Luton on lease to Yemen Airways for three months, and when this aircraft returned to Luton in March 1974, G-AVRM then left Luton as a replacement with Yemen. 1974, though, was a bad year for the British holiday airlines - inflation, the rise in the cost of fuel and other financial problems had caused there to be a sharp rise in the price of holidays, and consequently a drop in bookings for holidays abroad with all the major tour companies. Some airlines, including Britannia Airways were lucky and managed to lease some of their aircraft out to other airlines. Britannia Airways leased two Boeing 737s to Transavia of Holland, while the lease of the aircraft to Yemen Airways also continued. August 1974 saw the closure of Courtline and Donaldson, but Britannia Airways survived, and emerged at the end of the season as Britain's number one holiday airline.

Early in 1975 Britannia Airways realised that it would require to use the two 737s then on lease to Transavia. However, Transavia still required these aircraft, and consequently Britannia leased two Boeing 737s from United Airlines, and then sub-leased these aircraft to Transavia, thus releasing Britannia's own two aircraft which then returned to Luton. The two ex-United Airlines aircraft were N9059U and N9074U, but these aircraft never flew for Britannia Airways. During 1975, therefore, Britannia Airways had thirteen Boeing 737s in service, and these aircraft were flown on services from Birmingham, Bristol, Cardiff, Castle Donington, Gatwick, Glasgow, Luton, Manchester and Newcastle Airports. There were also occasional flights from Belfast Airport, and some ad-hoc charters from other British airports including Exeter. However, most of Britannia's work is now concerned with the carriage of holidaymakers from the United Kingdom to the Mediterranean, Greece, the Canary Islands, North Africa and Russia, although the airline still carries many thousands of serviceman, their wives and families from Luton to many Royal Air Force bases in north western Germany under Britannia's Rhine Army contract with the Ministry of Defence.

In just fourteen years, Britannia Airways has built itself up from an airline flying a small fleet of outdated piston engined Constellations on a handful of tour flights, to a major European holiday airline flying a large fleet of sophisticated and modern jet airliners on a network of international inclusive tour services almost too numerous to list.

Fleet:

Lockheed L049D Constellation	G-AHEN	c/n	1980	dd. 4.62,	wfu. 4.65.	
	G-ARVP	c/n	1967	dd. 4.62,	wfu. 4.65.	
Lockheed L049E Constellation	G-AHEL	c/n	1977	dd.10.62,	wfu. 12.64.	
	G-AMUP	c/n	2051	dd.10.62,	wfu. 8.65.	
Lockheed L149 Constellation	G-ARXE	c/n	1965	dd. 4.62,	wfu. 5.65,	painted in Skyways colours 1963.
Lockheed L749A Constellation	G-ALAK	c/n	2548	dd. 9.62,	sold 8.65.	
	G-ALAL	c/n	2549	dd. 9.62,	sold 8.65.	
	G-ANUR	c/n	2565	dd. 9.62,	sold 3.67.	
Bristol 175 Britannia 102	G-ANBA	c/n	12902	dd. 3.65,	wfu. 3.70.	
	G-ANBB	c/n	12903	dd.11.64,	written off	Ljubljana 1.9.66.
	G-ANBD	c/n	12905	lsd. 4.68,	ret. 10.68,	leased from B.K.S.Air Transport.
	G-ANBE	c/n	12906	dd. 2.66,	wfu. 12.70.	
	G-ANBF	c/n	12907	dd. 1.65,	wfu. 2.70.	
	G-ANBI	c/n	12910	dd. 4.66,	wfu. 9.69.	
	G-ANBJ	c/n	12911	dd. 5.66,	wfu. 12.70.	
	G-ANBL	c/n	12913	dd. 6.65,	wfu. 12.70.	
	G-ANBN	c/n	12915	lsd. 4.68,	ret. 10.68,	leased from Laker Airways.
	G-ANBO	c/n	12916	dd. 1.65,	wfu. 12.70.	
Boeing 707-373C	G-AYSI	c/n	18707	lsd. 2.71,	ret. 4.73,	leased from GATX/Boothe.
Boeing 707-355C	G-AYEX	c/n	19417	lsd.10.71,	ret. 5.73,	leased from Executive Jet Aviation
Boeing 737-204	G-AVRL	c/n	19709	dd. 7.68.		
	G-AVRM	c/n	19710	dd. 8.68.		
	G-AVRN	c/n	19711	dd. 4.69.		
	G-AVRO	c/n	19712	dd. 4.69.		
	G-AWSY	c/n	20236	dd. 5.69.		
Boeing 737-204C	G-AXNA	c/n	20282	dd. 3.70.		
	G-AXNB	c/n	20389	dd. 4.70,		'City of Birmingham'.
Boeing 737-2C4	G-AXNC	c/n	20417	dd. 5.70.		
Boeing 737-2C4A	G-BADP	c/n	20632	dd. 1.73.		
	G-BADR	c/n	20633	dd. 3.73.		
	G-BAZG	c/n	20806	dd. 2.74.		
	G-BAZH	c/n	20807	dd. 2.74.		

	G-BAZI	c/n 20808	dd. 3.74.	
Boeing 737-222	G-AZNZ	c/n 19074	dd. 3.72,	'City of Nottingham'.

BRITAVIA

Britavia Ltd. — Base: Blackbushe Airport.
British Aviation Services Ltd.
British Aviation Services (Malta) Ltd.

British Aviation Services was formed in November 1945 to act as technical advisors to the British Aviation Insurance Company. The company's first Managing Director was Air Commodore Griffith J.Powell who was formally an Imperial Airways pilot, and Senior Staff Officer of No.45 Group, Royal Air Force Transport Command at Montreal during the latter part of the War. The other members of the Board of directors of British Aviation Services were Mr.E.B.Ferguson (General Manager of the Pheonix Assurance Company, Chairman of the British Aviation Insurance Company, and Chairman of British Aviation Services); Sir Arthur Rogers (General Manager of the London and Lancashire Insurance Company, and Deputy Chairman of British Aviation Services); Sir James Dyer Simpson (Chief General Manager of the Royal Insurance Company and of the Liverpool and London and Globe Insurance Company); and Captain A.G.Lamplugh (underwriter and principal surveyor of the British Aviation Insurance Company).

British Aviation Services, thus, started operations purely as aviation consultants and advisers to airline operators and other aviation companies. However, this did not stop British Aviation Services from carrying out any flying - one of the first tasks was to ferry thirty six Dakotas across the Atlantic. These aircraft were flown from Royal Air Force Silloth, in Cumberland, to Cartierville near Montreal on behalf of Canadair, who then converted the aircraft for use by airlines as passenger transports. The ferrying of these aircraft took place between July 26 and December 11 1946, and British Aviation Services used its own crews for these flights. The company, at this time, employed almost one hundred people, and nearly all the crews were ex-Royal Air Force, with many coming from No.45 Group. This group carried out numerous trans-Atlantic ferry flights during the War, and, therefore, the crews were ideal for this job. After Canadair had completed work on the Dakotas, it was then British Aviation Services' responsibility to ferry the aircraft to their respective airline customers all over the world. Other ferry jobs carried out by British Aviation Services included the delivery of five Short Sandringhams to the Argentine in December 1945. British Aviation Services had been appointed as general agent to the Argentine Dodero Company, and this South American concern had purchased several types of British aircraft which British Aviation Services were then responsible for flying to Buenos Aires, including the Sandringhams. The company also ferried SABENA's new fleet of Douglas aircraft across the Atlantic during 1946 and 1947, and other aircraft were ferried from Montreal to airlines in Brazil, Ecuador, France and Sweden. Numerous aircraft were delivered to foreign governments all over the world,and this provided British Aviation Services with its main source of income at this time.

British Aviation Services set up its main British maintenance base at Blackbushe during 1946, and at this airfield the company employed some of the finest engineers available. A list of pilots employed by the company also made impressive reading - they included Captains D.W.Fenton, AFC., H.R.Banting and H.C.Graham, AFC., (who at one time held the record for a trans-Atlantic flight from Gander to Prestwick, when he flew a Mosquito over this route in five and a half hours while he was a member of No.45 Group, Royal Air Force). Air Commodore Powell's experience with No.45 Group had, therefore, enabled him to select some of the finest aircrew then available for the arduous task of trans-Atlantic ferry flights, and his knowledge was to give British Aviation Services great standing in the early post war years.

However, although the ferry business was booming in 1946 and 1947, it was realised that as soon as all the air forces and airlines had re-equipped with new aircraft, then the amount of ferrying work would die as quickly as it had grown. Therefore, British Aviation Services decided to expand its aviation consultancy side of the business - in addition to acting as British agent for the Argentine Dodero Company (later nationalised to become part of Flota Aerea Mercante Argentina), the company also acted as advisers to other organisations. British Aviation Services became the official adviser to the Government of Bermuda, and No.45 Group's Commanding Officer in Bermuda during the War, E.M.Ware, became British Aviation Services' representative in Bermuda.

British Aviation Services' first real experience of commercial airline operations came in October 1946 when Silver City Airways started long range passenger and freight charter operations with a fleet of Avro Lancastrians. Silver City Airways was an associate company of British Aviation Services, and was set up with the backing of British Aviation Services. Air Commodore G.J.Powell also became Managing Director of Silver City Airways, and several other employees of British Aviation Services also held positions in the hierarchy of Silver City Airways. However, Silver City's maintenance was initially undertaken at Langley Aerodrome, and while the technical side of Silver City Airways was completely separate from British Aviation Services, the administrative side of Silver City was looked after by its associate. Silver City Airways, though, was not British Aviation Services' only early post war experience of commercial airline operations - on January 1 1947 another airline company was founded named British Aviation Services (Malta) Ltd. Again the position of Managing Director was predictably filled by Air Commodore Powell, but a local manager was appointed to look after the day to day running of the company. Mr.A.Cassar had an office at Valetta, and he was responsible for the operation of this company, which undertook aircraft repair and servicing. British Aviation Services (Malta) also acted as aeronautical agents, and, with a fleet of three Airspeed Consuls, started services in January 1947 linking Luqa Airport, Malta with many Mediterranean ports. These ports included Araxos Athens, Catania, Naples, Rome, Tripoli and Tunis. In addition, many charters were undertaken, particularly on behalf of shipping companies, and the Consuls often plied their way back to the British Isles while engaged on charter work. This division in Malta employed three captains, led by Captain Bishop who acted as Chief Pilot, three radio officers and a number of ground engineers. In addition to caring for its own aircraft in Malta, British Aviation Services (Malta) also handled and serviced many of the aircraft of other British charter companies which passed through Luqa Airport.

By early 1947, therefore, British Aviation Services was engaged in a wide variety of aviation services worldwide. Although the ferrying side of the business was now dying, the air charter side was expanding, and through its associate company, Silver City Airways, British Aviation Services was to play an increasingly major role in post war charter operations. The two companies retained their separate identities, and on November 13 1947 one Bristol Freighter was flown to Karachi in Silver City Airways' markings to help in the evacuation of the Hindu refugees from Pakistan. One Dakota was also sent out to this area, and British Aviation Services set up its own maintenance facility at Karachi, and in addition to its own aircraft, it also services the Dakotas of Air Contractors and Kearsley Airways who were operating similar services to those of Silver City in Pakistan. The Bristol Freighter was stripped of all seats, and during its nine days of operations it carried one thousand, one hundred and five evacuees. The aircraft flew continually throughout the day to Ambata, Lahore, Lyallpur and Multan; and still operating within its all up weight of thirty seven thousand pounds, the Freighter carried as many as one hundred and fifty one passengers in one day. British Aviation Services' engineers were based in Pakistan, and as the aircraft flew non-stop throughout the day, all maintenance work had to be carried out during the night. This operation is just one example of the co-operation between British Aviation Services and Silver City Airways.

By 1948 the ferrying business had quietened considerably, but Dakotas owned by British Aviation Services often appeared at Prestwick on the first leg of their trans-Atlantic delivery flight, and over the next few years the overhaul, conversion and delivery of these aircraft provided British Aviation Services with a steady income. Another subsidiary company had been set up in Northern Rhodesia named Northern Rhodesian Air Services, and one Consul was operated by this company on air services within Rhodesia - Air Commodore Powell was also managing Director of this company.

In June 1949, though, changes came when the British Aviation Insurance Company decided to dispose of its interest in British Aviation Services, apart from the technical department of British Aviation Services which was retained by the Insurance Company to provide technical services for insurance interests. In order to achieve this reorganisation, another company, named Britavia, which had been formed some time before, had its share capital increased to two hundred thousand pounds to enable it to acquire a large proportion of the business carried on by British Aviation Services. Under this reorganisation, Britavia became the parent company of Silver City Airways and of the associated companies in Canada, Malta and Rhodesia. Air Commodore Powell became Managing Director of Britavia, and operations continued much as before. Apart from the aircraft of its subsidiaries, Britavia still operated no aircraft of its own.

In 1952 Britavia took over part of the premises formerly occupied by Westminster Aviation Servicing at Blackbushe, and from here Britavia carried out a large amount of maintenance work on behalf of many companies. In March 1953 Aquila Airways was bought out, although, like Silver City Airways, it continued to operate normally under its own name. Early in 1954 Britavia successfully tendered for a number of Air Ministry trooping contracts. In order to operate these services, Britavia bought a fleet of Handley Page Hermes from B.O.A.C. These aircraft were ideal for trooping services with their long range and relatively high capacity, and in April 1954 Britavia's Chairman, Mr.E.C.Mekie, announced that Britavia was to purchase six of these aircraft for use on its trooping services.

Unfortunately, though, all the Hermes needed work doing to them before they could leave Heathrow, and the B.O.A.C. employees refused to do this work as they felt that the sale of these aircraft to independent operators would lead to a reduction in the amount of their own work. The Hermes were, therefore, stranded at Heathrow and in the second week of May Mr.Mekie announced that Britavia intended to issue a writ against B.O.A.C. for breach of contract. Fortunately, this awkward situation was soon resolved, and the Hermes were flown to Blackbushe where the first aircraft entered service on the trooping runs in July 1954. Prior to the introduction of these aircraft, the P and O Shipping Group had bought a controlling interest in Britavia. During 1954 Britavia only undertook trooping flights to the Middle and Far East as far as Singapore, and the Hermes were also familiar sights in Egypt, the Persian Gulf, Cyprus and Malta carrying servicemen and their families. The initial success of these Hermes operations prompted Britavia to consider further expansion, and during the year the airline applied for permission to operate a scheduled all freight service from Blackbushe to Perth, Western Australia. Had this service been operated, Britavia would have used Hermes three times weekly in each direction with stops at Naples, Nicosia, Bahrein, Karachi, Colombo, Singapore and Sourabaya.

Over the next few years, the Hermes travelled to many parts of the world carrying troops - initially the aircraft flew these services with trooping serials, but later they adopted their normal civil registrations and Britavia's colour scheme. By 1957 the Hermes were also being used for many ad-hoc passenger and freight charter flights, including inclusive tour charter flights to Mediterranean destinations during the summer months. However, the Hermes rarely visited other airports within the British Isles, although on Christmas Eve 1958 two aircraft diverted into Stansted due to adverse weather conditions at Blackbushe, after carrying tourists to Paris for their Christmas holidays. In the spring of 1959 the trooping contracts came to an end, and the aircraft were, therefore, left without their regular work. The occasional ad-hoc charter was still flown - a load of tourists were bought from Dinard to Hurn on May 24 1959, but early in June the entire fleet of Hermes was moved to Manston Airport to inaugurate Silver City's new 'Silver Arrow' services to Paris. All the Hermes moved to this new base, and gradually they were all sprayed in Silver City's colour scheme, except for G-ALDX which was withdrawn at the end of the season. For the 1959 season, the Britavia Hermes operated many inclusive tour flights from Manston, and the aircraft also appeared at Gatwick on charter work - G-ALDU flew a Gatwick to Berlin charter flight on July 29, returning to Gatwick on August 2. Another Hermes (G-ALDX) flew a service from Gatwick to Lisbon on August 20, while G-ALDI flew a service Gatwick - Speke - Gatwick on July 18. The aircraft made several other visits to Gatwick - mostly on charter flights to European destinations with holidaymakers, although the aircraft also diverted to Gatwick when Manston's weather conditions would not permit a landing there.

By the end of 1959 all the Hermes were in Silver City's colours, and the name of Britavia disappeared from the aircraft. However, Britavia remained as the holding company for Silver City Airways until early in 1962, when the company sold its shares to the Air Holdings Group and Britavia's name finally disappeared. At the time of its take-over, Britavia's board of directors consisted of Mr.Eoin C.Mekie (Chairman), Sir Donald F.Anderson, Sir Arthur V.Harvey, Air Vice-Marshall S.D.Macdonald, General Sir Edwin L.Morris, Sir Edward Wilshaw, Mr.C.M. Fox, Mr.R.G.Grout, Mr.A.J.Maitland-Makgill-Chrichton and Mr.A.E.Barnes.

Fleet:

Beech C 18S	G-AKCZ	c/n 6116	dd. 1.49,	sold	9.50.	
Airspeed AS65 Consul	G-AHRK	c/n 3096	dd. 7.46,	sold	4.52,	operated by Northern Rhodesian Aviation Services.
	G-AIBF	c/n 3422	dd. 8.46,	wfu.	4.54,	operated by B.A.S.(Malta) Ltd.
	G-AIKY	c/n 4324	dd.12.46,	sold	4.47,	operated by B.A.S.(Malta) Ltd.
	G-AJGH	c/n 5124	dd. 5.47,	sold	9.47,	operated by B.A.S.(Malta) Ltd.
	G-AJNG	c/n 5146	rd.11.48,	sold	9.49.	
Douglas C47B Dakota	G-AGNG	c/n 26998	dd. 1.51,	sold	1.51.	
Douglas C47A Dakota	G-AIRG	c/n 25288	dd.11.46,	sold	3.48.	
Douglas C47B Dakota	G-AKAR	c/n 26889	dd. . ,	sold	12.50.	
Lockheed 14	G-AGBG	c/n 1421	dd. .46,	sold	1.47.	
Handley Page 81 Hermes 4	G-ALDI	c/n HP81/10	dd. 6.54,	ISOM	,	trooping as XJ309.
	G-ALDJ	c/n HP81/11	dd. 5.54,	written off		Blackbushe 6.11.56.
	G-ALDK	c/n HP81/12	dd. 6.54,	written off		Karachi 5.8.56, trooping as XJ281.
	G-ALDP	c/n HP81/17	dd. 6.54,	ISOM	,	trooping as XJ269.
	G-ALDU	c/n HP81/21	dd.12.56,	ISOM	,	trooping as XJ280.
	G-ALDX	c/n HP81/24	dd. 1.57,	wfu.	1.60,	trooping as XJ267.

BAF

British Air Ferries Ltd.
British United Air Ferries Ltd.

Bases: Southend and Lydd Airports.

When the vehicle ferry division of Silver City Airways and Channel Air Bridge were brought together in January 1962, it became apparent that the two separate divisions would have to be amalgamated, and consequently on January 1 1963, British United Air Ferries officially came into being. By this time Channel Air Bridge, operating from Southend, flew both short cross-Channel services and also longer range services to Basle, Geneva and Strasbourg. Silver City's operations, on the other hand, were from Lydd, Bournemouth and Southampton with most flights being just short hops across the Channel. After the amalgamation, British United Air Ferries maintained Southend and Lydd as its main bases, with several Bristol Freighters based at Southampton. From Lydd, the airline operated in association with Cie. Air Transport, with services to Le Touquet, Ostend and Rotterdam. The Carvairs were initially flown from Southend, and two aircraft were equipped to carry up to fifty five passengers and three cars on services from Southend to Ostend. At the time of its formation, British United Air Ferries was at the height of its vehicle ferry operations, and during the first two weeks of August 1963, the traffic on the cross Channel services exceeded all previous records. The airline used twenty nine aircraft during this period, carrying 12,150 cars and 51,269 passengers on 4,762 Channel crossings - an average of 320 a day. The daily average load was 810 cars, and 3,400 passengers. During the summer of 1963, all services carried increased traffic, with the most spectacular growth being on the deep penetration routes to Basle, Geneva and Strasbourg, and this led to the introduction of a service to Liege from Lydd and Southend in 1964. Prior to this, one of the airline's Carvair (G-ASDC) made the first Carvair landing at Lydd on December 13 1963.

Apart from the introduction of the Liege service in 1964, British United Air Ferries also started a route between Coventry and Calais on May 2 1964 using Bristol Freighters, and on October 31 1964 G-ASDC operated the first Carvair Jersey to Hurn service. During 1965, the airline's Carvairs flew many ad-hoc charters, and one of their Carvairs was fitted with eight fuel tanks instead of six, giving it a range advantage over the other aircraft in the fleet. This aircraft (G-APNH) left Lyneham on January 11 1965 on the first British Government charter flight to the Woomera Rocket Range in Australia, but enroute the rocket, which the aircraft was carrying, cracked due to severe heat, and at San Francisco the aircraft was forced to turn back for the United Kingdom. However, although the Carvairs were never contracted to carry out any more flights of this nature, the aircraft were used for further Government contracts to the Mediterranean and Middle East. A further unusual use for the Carvair came in December 1965 when, because of the Rhodesian declaration of UDI in November, aircraft were needed to transport essential supplies from Dar-es-Salaam to Lusaka, the Zambian capital. However, British United were unwilling to let their name be connected with this operation, and consequently two Carvairs were painted up in Air Ferry colours, with G-ASKG leaving Southend on December 28, and G-APNH following two days later.

During 1966, the vehicle ferry service continued from Coventry to Calais, but in June 1966 the joint freight service with SABENA between Southend from Ostend came to an end, and British United Air Ferries continued to operate this service alone. During the winter of 1966, the Coventry to Calais services was flown via Southend, but on February 28 1967 this service was axed by the airline. At this time, the company also decided to suspend all its deep penetration services, and also other uneconomic services. These were from Manchester to Calais; Coventry to Calais; Southend to Basle and Geneva; Gatwick to Le Touquet; and Lydd to Basle and Geneva. This cut back meant that many aircraft were put into storage or withdrawn from use altogether. On March 31 1967, Bristol Freighter G-AMLP flew the last Bristol Freighter service from Southend when it flew to Calais and return. Subsequently, the Freighters only appeared at Southend for maintenance, when they flew charters, or when loads on normal Carvair services were heavy, and they could be used as back-up aircraft. During April 1967, the long range Carvair (G-APNH) was operated in the Trucial States, Muscat and Oman on behalf of Shell Petroleum, carrying oil drilling equipment around the Middle East.

The Bristol Freighter remained in service to operate the services from Lydd, but G-ANMF which was the airline's only short nosed series 31 was withdrawn in August 1967. This aircraft was equipped with 51 seats, and was used to fly many of the airline's rail-air services. British United had agreements with several railway companies in France, Belgium, Britain and Holland whereby passengers could travel from London to Paris, Brussels and Amsterdam by a combination of rail and air services. These rail-air services still continue today, and are particularly popular in the summer time. In addition to the rail-air aircraft, British United Air Ferries also operated one aircraft (G-AOUV) for horse charters.

On October 1 1967, British United Air Ferries changed its name to British Air Ferries, and at the same time it became a completely separate division, in the Air Holdings Group, from British United Airways. However, by this time the airline was beginning to feel the severe competition from the sea ferries, and British Air Ferries began to concentrate more and more on the carriage of cargo and passengers, as opposed to vehicles. This then led to the gradual rundown of the vehicle ferry services, particularly those from Lydd, leading to the last Lydd to Le Touquet service being flown by Bristol Freighter G-ANWM at the end of October 1970. However, Lydd continued to be connected with Ostend by Carvairs, but this service, too, had ended by early 1971. For the 1970 summer season, the airline operated its first all passenger aircraft in the form of a leased Aer Lingus Viscount, and this was used on services from Southend to Ostend and Le Touquet. During 1971, this aircraft and two leased HS748s, were used on these services, and during the summer of 1973 an Alidair Viscount was used. The first HS748 passenger service was flown on March 11 1971 when G-ATMJ flew from Southend to Ostend and return. During 1971 the 748s were flown on a weekly flight from Norwich to Esbjerg on behalf of an oil company.

However, on October 27 1971 the airline was bought out by Mr.T.D.Keegan of Trans Meridian fame and this change of ownership soon became apparent in British Air Ferries' operations. Apart from an application in British Air Ferries' name for a weekly cargo service from Stansted to Rhodesia using CL-44s, the airline reopened its vehicle ferry, passenger and cargo service from Southend to Basle on April 1 1972. This service was also flown from Stansted, and numerous aircraft were chartered during the winter of 1971 and summer of 1972 - these aircraft included Pomair DC6Bs, British Island Airways Heralds, Air International Viscount, Delta Air Transport DC6Bs, Dan-Air 748s, Northeast Viscounts and Invicta Vanguards. On January 13 1972, Carvair G-AOFW made a proving flight between Guernsey and Castle Donington on behalf of a consortium of Guernsey businessmen who had hoped to operate a five times a week service between the two points. However, this service never got off the ground, although services were started from Jersey and Guernsey to Bournemouth carrying cargo and passengers, and on March 19 Carvair G-ASDC inaugurated a Coventry to Jersey and Guernsey schedule. These services were for both passengers and cargo, and it had been hoped to open further services to Ostend and Le Touquet from Coventry, but these routes were never started, and all Coventry services were suspended at the end of the 1973 summer season. In March 1972, one of the Trans Meridian CL-44s was transferred to British Air Ferries in an attempt to modernise the airline's ageing fleet,and on March 28 1972 the CL-44 (G-ATID) operated its first service from Southend to Ostend, carrying a total of 170 passengers. Flights from Stansted were also inaugurated over the Easter period, using leased aircraft, and these were flown to Basle, Ostend and Rotterdam.

On May 13 1972, the second CL-44 entered service in a passenger/vehicle configuration. However, a third CL-44 failed to enter service when operating problems with the type caused all flights to be stopped from Southend, and due to a lack of passengers all flights from Stansted ceased on June 30. By this time more Carvairs had entered service, and the CL-44s were handed back to Trans Meridian. The Basle service was now flown thrice weekly from Southend, and during 1973 British Air Ferries entered into an agreement with Intra Airways for the joint operation of the freight services between the Channel Islands and Bournemouth. The aircraft are also frequently used for ad-hoc charter work, and this includes many flights on behalf of the Ford Motor Company from West Germany to Southend and Liverpool carrying spare parts. The Carvairs are also ideal aircraft for carrying outsize loads, and in this capacity they occasionally visit Aberdeen in connection with the oil rig support work being carried out from there. Late in 1974, a dock strike in Northern Ireland caused the airline's Carvairs to fly numerous services during the night between Belfast and Southend carrying cigarettes.

In January 1975, British Air Ferries completed the purchase of three Eastern Provincial Airways' Heralds, and the first aircraft arrived at Southend on January 28, and after a quick respray, started crew training on February 4, and on February 20 a service was flown from Southend to Rotterdam. Crew training was carried out with the help of British Island Airways, and by the end of July, all three aircraft had arrived from Canada, and had replaced the Carvairs on some of the services to Le Touquet, Ostend and Rotterdam from Southend. As the Carvairs were released from these routes, they were converted to an all cargo configuration, and with its paint stripped off 'Big Louie' became 'Plain Jane' during April and continues in service as an all-cargo aircraft. For the future, British Air Ferries hopes to introduce futher long range services from Southend to Bordeaux, Marseilles, Milan and Nimes.

During 1975 the Carvairs still remained in service on the route from Southend to Rotterdam but during the year many ad-hoc charters were undertaken. One aircraft (G-AOFW) was used for transporting 'pop groups' and their equipment around Europe, and for this purpose the aircraft acquired a 'Startrekking 75' emblem on the fuselage. Among the ad-hoc charter work undertaken was a flight from Humberside to Groningen in Holland by G-AXAI on Monday September 23 - this was the first visit of a Carvair to the newly opened airfield at Humberside, and on Tuesday September 9 1975 G-AOFW operated an oil rig charter flight from Norwich to Pescara in Italy

One of Bond Air Services' many Halifaxes which operated on the Berlin Airlift.

A J Jackson

BEA Airtours' Comets rarely visited British airports apart from Gatwick, but G-ARJL was captured at Newcastle on October 19 1970.

Ian MacFarlane

British Airtours' smart Boeing 707 G-APFF taxies in at Gatwick after completing another inclusive tour flight.

Tony Merton Jones

G-AIAR operated two hundred and forty seven sorties on the Berlin Airlift in British American Air Services' colours, and then continued charter flying until late in 1950.

A J Jackson

One of the smallest of all post-war charter operators, British and Continental used one Anson for ad-hoc charter work from Southend.

J Jackson

BR2001, alias G-AWWD, cleans up after take-off at Gatwick for Los Angeles on September 29 1975, to begin its climb to its cruising level.

Tony Merton Jones

The second British airline to introduce BAC 1-11's was British Eagle, and here G-ATPJ is seen shortly before touching down.

British Aircraft Corporation

In Gatwick's early jet days, British United's 1-11's dominated most parts of the apron - G-ASJF, a 1-11 srs.201 is seen in this photograph.

British Aircraft Corporation

British United's second VC-10, G-ASIX, is caught here at an impressive moment during take-off.

British Aircraft Corporation

- the return trip on the following day terminated at Southend instead of Norwich. Other charter flights have been flown by the Heralds - on Sunday mornings in August and September 1975 the Heralds flew several charters from Zurich to Edinburgh.

The all cargo Carvair operates freight services to many of British Air Ferries' normal scheduled destinations, and during 1975 this aircraft also visited many other European airports including Dusseldorf, Naples, and Stuttgart.

British Air Ferries' operations are now, therefore, a mixture of passenger and freight, scheduled and charter operations within Europe. Although the Carvairs are still giving excellent service, a replacement will have to be found shortly to take over from these elderly aeroplanes, particularly in the cargo role.

Fleet:

Bristol 170 Freighter Mk.21	G-AHJI	c/n 12741	ISOF	wfu. 11.63,	'City of Bath'.
	G-AIFM	c/n 12773	ISOF	wfu. 11.63,	'City of Carlisle'.
	G-AIME	c/n 12795	ISOF	wfu. 10.63,	'City of Exeter'.
Bristol 170 Freighter Mk.31	G-ANMF	c/n 13216	ISOF	wfu. 8.67,	'Victory'.
Bristol 170 Freighter Mk.32	G-AMLP	c/n 13078	ISOF	sold 11.70,	'Vanguard'.
	G-AMSA	c/n 13142	ISOF	wfu. 6.65,	'Voyager'.
	G-AMWA	c/n 13073	ISOF	written off Guernsey 24.9.63,	'City of London'.
	G-AMWB	c/n 13127	ISOF	wfu. 6.67,	'City of Salisbury'.
	G-AMWC	c/n 13128	ISOF	wfu. 5.65,	'City of Durham'.
	G-AMWD	c/n 13131	ISOF	wfu. 1.66,	'City of Leicester' later 'City of Hereford'.
	G-AMWE	c/n 13132	ISOF	wfu. 6.66,	'City of York'.
	G-AMWF	c/n 13133	ISOF	wfu. 11.67,	'City of Edinburgh'.
	G-ANVR	c/n 13251	ISOF	sold 3.71,	'Valiant'.
	G-ANVS	c/n 13252	ISOF	wfu. 11.67,	'Vigilant'.
	G-ANWG	c/n 13211	ISOF	wfu. 3.68,	'City of Winchester'.
	G-ANWH	c/n 13212	ISOF	wfu. 6.69,	'City of Hereford'.
	G-ANWI	c/n 13213	ISOF	wfu. 3.68,	'City of Glasgow'.
	G-ANWJ	c/n 13254	ISOF	wfu. 8.70,	'City of Bristol'.
	G-ANWK	c/n 13259	ISOF	wfu. 10.69,	'Quatorze Juillet'.
	G-ANWM	c/n 13261	ISOF	wfu. 8.70,	'City of Aberdeen'.
	G-ANWN	c/n 13262	ISOF	wfu. 8.70,	'City of Hull'.
	G-AOUU	c/n 13257	ISOF	wfu. 6.65,	'Venture'.
	G-AOUV	c/n 13258	ISOF	wfu. 5.67,	'Valour'.
	G-APAU	c/n 13256	ISOF	sold 4.71,	'Versatile' later 'City of Edinburgh'.
	G-APAV	c/n 13263	ISOF	sold 10.70,	'Viceroy'.
ATL.98 Carvair	G-ANYB	c/n 01	ISOF	wfu. 8.70,	'Golden Gate Bridge'.
	G-AOFW	c/n 12	dd. 3.65		'Big John'.
	G-APNH	c/n 11	dd. 1.65,	wfu. 3.71,	'Menai Bridge'.
	G-AREK	c/n 05	dd. 4.72,	sold 6.74,	'Porky Pete'.
	G-ARSD	c/n 02	ISOF	wfu. 8.70,	'Chelsea Bridge'.
	G-ASDC	c/n 07	dd. 3.63		'Pont du Rhin' later 'Big Louie' later 'Plain Jane'.
	G-ASHZ	c/n 09	dd. 7.63		'Massburg' later 'Fat Annie'.
	G-ASKD	c/n 14	dd. 8.73,	sold 11.74,	'North Sea Viking Express'.
	G-ASKG	c/n 10	dd. 2.64,	sold 2.75,	'Channel Bridge' later 'Big Joe'.
	G-ASKN	c/n 13	dd. 2.64		'Pont D'Avignon' later 'Big Bill'.
	G-ATRV	c/n 15	dd. 3.66,	sold 5.67.	
	G-AXAI	c/n 17	dd. 4.69		'Fat Albert'.
HS.748 Series 222	G-ATMI	c/n 1592	lsd. 4.71,	ret. 11.71,	leased from Courtline.
	G-ATMJ	c/n 1593	lsd.11.70,	ret. 10.71,	leased from Courtline.
Canadair CL-44D-4	G-ATZH	c/n 21	dd. 5.72,	transferred	to Trans Meridian 7.72.
	G-AZIN	c/n 19	dd. 3.72,	transferred	to Trans Meridian 7.72.
Douglas DC 4	G-BANO	c/n 42988	dd. 2.73,	wfu. .73,	never entered service.
	G-BANP	c/n 42951	dd. 1.73,	sold 2.74.	
Vickers 803 Viscount	EI-AOI	c/n 179	lsd. 3.70,	ret. 9.71,	leased from Aer Lingus.
Vickers 812 Viscount	G-AVIW	c/n 358	lsd. 4.73,	ret. 10.73,	leased from Alidair.
Handley Page Herald 202	G-BCZG	c/n 159	dd. 3.75,		'Rupert Keegan'.
Handley Page Herald 206	G-BCWE	c/n 166	dd. 1.75,		'Jeremy Keegan'.
	G-BDFE	c/n 167	dd. 7.75,		'Rory Keegan'.

BRITISH AIR SERVICES

British Air Services Ltd.
B.K.S. Air Transport Ltd.
Cambrian Airways Ltd.

Bases: as for B.K.S. Air Transport and Cambrian Airways.

British Air Services was formed by British European Airways in March 1967 to look after its financial interests in two British independent airlines - B.K.S. Air Transport and Cambrian Airways. At the time of British Air Services' formation, B.E.A. held a fifty per cent interest in B.K.S. (acquired in November 1966) and a thirty three per cent interest in Combrian (acquired in 1958). Chairman of British Air Services was Mr.A.C. Ping, and the board of directors mainly consisted of members of both B.K.S. and Cambrian boards, together with several B.E.A. representatives. In November 1967, both Cambrian and B.K.S. Air Transport became wholly owned subsidiaries of British Air Services,although both companies maintained their separate identities until becoming part of British Airways in 1972.

During its comparatively brief existence, British Air Services carried out an evaluation of operating local air services within the British Isles, and it is probably for this evaluation that British Air Services is best known. The three aircraft considered for use on these services were the Saunders ST-27, the de Havilland Twin Otter and the Short Skyliner. The first two aircraft were demonstrated to British Air Services at several airports already used by the airline, while the Skyliner undertook a thorough evaluation over the routes proposed by British Air Services. The Skyliner (G-ASZJ) made a tour of the British Isles in September 1970 jointly sponsored by Short Brothers and British Air Services. This tour lasted for five days and took in twenty one cities, and was timed to take place after the 1970 Farnborough Air Show.

Thus at half past eight on the morning of Monday September 14 1970, the Skyliner took off from Heathrow for the first stage of its tour around the British Isles, landing at Cambridge just over half an hour later. At Cambridge the aircraft undertook demonstration flights, and a Press Conference also took place. From Cambridge, the Skyliner flew onto Peterborough, then to Norwich, to Hull and finally to Leeds where the aircraft then night stopped. On the following day, a Press Conference took place at Leeds, before the aircraft departed for Teesside, and then on from Teesside to Blackpool and then to Newcastle, where the aircraft then rested for the night. On Wednesday morning G-ASZJ left Newcastle for Edinburgh, and then it flew onto Abbotsinch, Prestwick and Carlisle before rounding off the day with a flight to Liverpool. During the last two days of the tour, the Skyliner visited many airfields in the South West of England and South Wales. On Thursday morning the aircraft took off from Liverpool in mid-morning arriving at Birmingham half an hour later. From Birminghan, the Skyliner continued its tour with a flight to Heathrow, then from Heathrow onto Swansea, and from Swansea to Cardiff. The last day of the tour started early in the morning with a flight from Cardiff to Exeter at nine o'clock in the morning, followed by flights to Plymouth, Southampton, Bristol and finally to Heathrow, where the Skyliner then completed its demonstration tour. At every airfield on the tour the Skyliner had carried out a series of short demonstration flights with the exceptions of Prestwick and Heathrow. Press Conferences, though, were held at all airports visited by the Skyliner.

On each of these flights, the Skyliner carried up to nineteen passengers - although the aircraft was equipped with twenty two seats, only nineteen passengers could be carried as Shorts had not worked out the necessary stress calculations for the rearward facing row of four seats behind the flight deck. Thus, eighteen passengers were accommodated in the passenger cabin, and the nineteenth passenger had the pleasure of sampling third level airline operation at first hand, by sitting in the right hand seat beside the captain. At the time, though, it was pointed out that if these services were to be flown without a cabin attendant, then the maximum number of passengers that could be carried legally would be nineteen. After the demonstration tour, the eight hundred people who had sampled this new type of service were asked to complete a questionaire, and to raise any points related to the type of operation that they personally would like to see undertaken by British Air Services.

During the period of evaluation, British Air Services put forward several different ideas for the types of services to be operated. The first idea was for a single hub to be based at Birmingham with services radiating to Bristol/Exeter, Southampton, Heathrow, Hull, Manchester/Newcastle, Leeds, Blackpool and Liverpool. The second idea was for a series of hubs with a complicated network of services linking the many towns with these hubs. One hub would have been at Cambridge with services to Norwich, Hull, Sheffield, Birmingham, Heathrow/ Southampton, and Bristol; a second hub at Birmingham would have been linked to Southend, Heathrow, Southampton, Liverpool, Blackpool, Manchester/Newcastle, Leeds and Hull; a third hub at Bristol would have had services to Southampton, Bournemouth, Exeter/Plymouth, Cardiff, Swansea, Liverpool, Wolverhampton, Birmingham and an air-strip at Surrey Docks in London; the last hub at Sheffield would have served Newcastle, Teesside, Hull, the Surrey Docks, Birmingham, Wolverhampton, Bolton, Preston and Carlisle. These routes were,of course, only a very basic plan of what the businessman could have expected, and they would have been operated subject to approval from the appropriate licensing authorities.

The idea of a multi hub system would have probably have required the purchase of up to six commuter air-craft of the Skyliner size - the idea of a single hub might have required a smaller number of aircraft, and although this network would have been less extensive than the multi hub system, British Air Services tended to favour the single hub idea. Had these services ever started, the most likely services to have been operated initially would have been Birmingham to Heathrow (seven return trips per day), Southampton (seven), Bristol and Exeter (three), Liverpool (two), Manchester and Newcastle (four), Leeds (six) and Hull (two).

This was the first real attempt by any British airline to set up a comprehensive network of scheduled third level services within the British Isles. Although these services never started, several other British airlines have since developed many services similar to those proposed by British Air Services. These airlines include Air Anglia and Dan-Air Services who, although using rather larger aircraft, now link many of Britain's provincial cities with scheduled services. British Airways did later purchase two Short Skyliners for use on their services in Scotland, but the aircraft did not prove to be particularly suitable for British Airways' operations, and the aircraft were sold.

No fleet list is included for British Air Services as its aircraft were operated by B.K.S. Air Transport and Cambrian Airways, and thus the aircraft are included in their fleet lists.

BRITISH AIRTOURS

British Airtours Ltd.
B.E.A. Airtours Ltd.

Base: London (Gatwick) Airport.

Formed as a subsidiary of British European Airways on April 24 1969 as B.E.A. Airtours, this company started inclusive tour charter operations from Gatwick with a Comet flight to Palma on March 5 1970. For the summer season the airline had a total of nine Comet 4Bs in service, all obtained from the parent company. During 1970 the airline flew services on behalf of many tour operators, but all departures apart from one were from Gatwick Airport, where the airline had acquired hanger space in which to set up a maintenance division. The one non-Gatwick departure was a weekly Birmingham to Palma service - the first service was flown by Comet G-ARJN on May 16. The airline stated, during 1970, that it hoped to corner twenty per cent of the inclusive tour charter market by the following summer. During 1970 from Gatwick, B.E.A. Airtours flew holidaymakers to Alicante, Gerona, Ibiza, Malaga, Nicosia, Palma, Pisa, Rimini, Tenerife, Trieste and Venice. The first summer went well for the company, and by the end of the season B.E.A. Airtours was looking around for larger aircraft for its operations, and in November 1970 B.E.A. Airtours announced that it would be buying a fleet of seven Boeing 707-123Bs from American Airlines at a total coast of £10.4m. This became the subject of a fierce political argument, with the corporations wanting Airtours to purchase surplus B.E.A. and B.O.A.C. aircraft, and the contract with American Airlines was cancelled.

While Northeast Airlines were having their Tridents painted up in their new colour scheme, several Airtours Comets were leased to Northeast for their London to Newcastle services, and also for their inclusive tours from Newcastle on behalf of Airway Holidays. During this time the aircraft carried dual B.K.S. and Northeast markings in the cabin, and this operation lasted from October 13 until November 18 1970. The first aircraft used being G-ARJN, and the last aircraft G-ARJL. During the winter months the Comets were used extensively by B.E.A. on many of their scheduled services from Heathrow, and only a small number of Airtours' Comets remained at Gatwick for use on the inclusive tour charters. The summer of 1971 saw little change from the previous summer, except for the addition of one Comet to the fleet, and the ending of the Birmingham to Palma service. In December, the first of many ex-B.O.A.C. Boeing 707s arrived, and by the summer of 1972 two aircraft were in service flying inclusive tours from Gatwick. All the Comets remained for the 1972 season, and the 707s were used on inclusive tours from Newcastle in addition to those from Gatwick. During the winter of 1972 two further 707s joined the fleet, and B.E.A. Airtours also used the aircraft for trans-Atlantic passenger charters to North America during the winter months.

1973 was to be the last season for the Comet with only five aircraft left in service, and when the last Comet arrived at Gatwick Airport from Le Bourget on October 31, the total number of passengers carried by the airline's Comets was two million. During the summer of 1973 the 707s were used on fortnightly services from Newcastle to Athens and Corfu. Seven Boeing 707s were now in service, and these aircraft were flying inclusive tours to all parts of Southern Europe and North Africa, including the Canary Islands, Morocco, Tunisia and Greece. Two of the 707s flew scheduled services on behalf of Syrian Arab Airlines during 1974, but this contract has now ended, and most of the airline's operations are still centred on Gatwick. The company, like its parent, now trades as British Airtours and the 707s display the same colours as those aircraft of British Airways. During 1975 British Airtours' Boeing 707s were used on many of British Airways' flights to Malta including those departures from London and Manchester.

Fleet:

DH106 Comet 4B	G-APMC	c/n	6423	dd. 2.71, sold 10.73.
	G-APMD	c/n	6435	dd. 3.70, sold 9.72.

DH106 Comet 4B	G-APME	c/n	6436	dd. 8.70,	sold 4.72.
	G-APMF	c/n	6426	dd. 4.70,	sold 1.73.
	G-APMG	c/n	6442	dd. 3.70,	sold 4.73.
	G-ARCP	c/n	6451	dd. 5.70,	sold 10.73.
	G-ARGM	c/n	6453	dd. 5.70,	sold 10.73.
	G-ARJK	c/n	6452	dd. 3.70,	sold 10.73.
	G-ARJL	c/n	6455	dd. 3.70,	sold 10.73.
	G-ARJN	c/n	6459	dd. 3.70,	sold 2.73.
Boeing 707-436	G-APFB	c/n	17707	dd. 3.74.	
	G-APFD	c/n	17705	dd. 2.73.	
	G-APFF	c/n	17707	dd. 5.74.	
	G-APFG	c/n	17708	dd. 3.73.	
	G-APFH	c/n	17709	dd. 1.72,	sold 3.75.
	G-APFI	c/n	17710	dd. 1.75.	
	G-APFK	c/n	17712	dd.12.71.	
	G-APFL	c/n	17713	dd.12.72.	
	G-APFO	c/n	17716	dd.11.72.	
Boeing 707-465	G-ARWD	c/n	18372	dd. 1.73.	

BRITISH AIR TRANSPORT

British Air Transport Ltd.

Bases: Redhill and Kenley Aerodromes, and Croydon Airport.

Formed originally in 1932, British Air Transport resumed air charter operations from Kenley Aerodrome in May 1946 with two Austers and an Airspeed Consul. These aircraft were used on passenger and freight charters, pleasure flying and aerial photographic work from both Kenley and Croydon. The company opened an office at Croydon and appointed Mr.Maurice Houdret as their traffic manager to look after their charter operations from Croydon. In October 1946 three more Consuls were acquired, and these aircraft were used on several passenger charter flights from Croydon to the Continent. At Kenley, British Air Transport undertook the maintenance and overhaul of many aircraft, and in May 1947 the airline reopened the Redhill Flying Club. British Air Transport owned the airfield at Redhill, and as soon as the military authorities had departed from the airfield after the end of the war, British Air Transport set to work to make Redhill available for civilian flying. At Redhill the company set up its own overhaul facilities, and the base at Kenley was closed. Also at Redhill the company leased out hanger space to other companies who wished to maintain their aircraft there, including Southern Aircraft (Gatwick) Ltd.

The Redhill Flying Club started flying instruction, once again, in May 1947 with four Auster Autocrats and three Miles Magisters. All these aircraft were used for dual flying instruction, and the rates varied from three pounds an hour solo to three pounds and ten shillings an hour dual. The club's Chief Flying Instructor was Squadron Leader J.L.Hubbard, and one of the Autocrats (G-AHCK) was often used for tugging duties by a local gliding club as it was equipped with a glider towing hook. May was quite a month for British Air Transport - on May 25 the sole airworthy de Havilland Flamingo was delivered from Gatwick to Redhill for operation by British Air Transport. This aircraft had been overhauled by Southern Aircraft at Gatwick, but it did not enter service, though, until early June, and throughout the summer the aircraft operated passenger and freight charters from Croydon. During the summer of 1947 the Consuls were familiar visitors to the Channel Islands and to the Isle of Wight. In addition, these aircraft would often be seen at many of the airfields in Northern France and Belgium, and services were flown from Croydon to many racemeetings around the country. During the summer, British Air Transport opened two daily newspaper services from Croydon to the Channel Islands and Paris. These services were operated under contract to the press, and in order to operate them, a fleet of former Royal Air Force Avro Ansons was purchased. These aircraft were converted by British Air Transport at Redhill, and the first Anson entered service during August 1947. The Consuls were also used on these services, and outbound to Paris the aircraft carried one thousand one hundred pounds of English newspapers, while on its return flight the aircraft brought back eight hundred and eighty pounds of the 'New York Herald Tribune'. The aircraft which flew to Guernsey and Jersey also carried English newspapers outbound, but on its return journey the aircraft came back empty. These services were flown in the early hours of the morning, and at weekends when the loads were particularly heavy, then the company would use the Flamingo on these services.

Throughout the winter of 1947, therefore, British Air Transport was able to keep its aircraft busy flying newspapers from Croydon to Paris and the Channel Islands. However, one of the Ansons was lost on November 20 1947 when it crashed on the Isle of Wight after a navigational error, resulting in ths loss of the crew and the Channel Islands' morning newspapers. Although British Air Transport's Flamingo was not especially active, the airline decided that it would like to obtain more of these aircraft. In October 1946 Tiltman Langley Laboratories had moved into premises at Redhill, and by this time the company had gained a wealth of experience in aircraft construction, and therefore British Air Transport asked Tiltman Langley to rebuild two de Havilland Flamingos. British Air Transport had purchased four of these aircraft, but initially only two were to be rebuilt. Even though the Flamingo was a pre-war design, the aircraft was still able to meet the latest I.C.A.O. requirements for performance after the failure of one engine. This fact, coupled together with the Flamingo's good profit earning capacity and airfield performance, led to its selection as British Air Transport's replacement for its Consuls and Ansons. The work on these aircraft continued until the end of 1948, but was then suspended when British Air Transport decided that they no longer wished to operate these aircraft.

During 1948 British Air Transport was one of the first companies to be granted a B.E.A.C. Associate Agreement when the airline had its Croydon to Cowes scheduled service approved. On this service, the airline operated mainly its Consul aircraft, but the Ansons also appeared occasionally, although many had been converted to freighter configuration for their newspaper runs. The flying club's activities were going well, and at the end of its first year of post war operations in May 1948, the Redhill Flying Club had flown over two thousand hours. However, towards the end of 1948 the newspaper contracts started to be passed to Transair, and by early 1949 the airline was beginning to be run down. The service to Cowes on the Isle of Wight was again flown during the 1949 summer season, and gradually the airline's operations were run down. Many of the Ansons were sold to Transair for use on their newspaper services, and the Consuls were also sold. The Flamingo, for all its beauty, had flown little, and by the end of 1951 this aircraft had lapsed into disuse, and after sitting at Redhill for several years, the aircraft was, sadly, broken up in 1954. All air transport operations came to a close at the end of 1951, but British Air Transport continued in existence as operators of the Redhill Flying Club who were still flying a fleet of Austers and Magisters.

Today, Redhill Aerodrome has changed little, and is still owned and operated by British Air Transport, but the sound of the Anson's Cheetahs has long gone, and the airfield is now better known as the home of Bristow Helicopters and the Tiger Club.

Fleet:

Auster J/1 Autocrat	G-AHAY	c/n	1956	dd. 4.46,	ISOC.
	G-AHCK	c/n	1973	dd. 5.46,	ISOC.
Airspeed AS65 Consul	G-AHEH	c/n	3362	dd. 5.46,	sold 11.48.
	G-AHFS	c/n	2942	dd. 6.46,	ISOC.
	G-AIDY	c/n	3094	dd.10.46,	written off Berck, France 14.6.48.
	G-AIDY	c/n	4404	dd.10.46,	sold 7.51.

Airspeed AS65 Consul	G-AIEA	c/n	4320	dd.10.46, sold 10.50.
Avro 652A Anson 1	G-AHKH	ex.	EG526	dd. 1.48, ISOC.
	G-AIWW	ex.	MG569	ca.11.47, written off Isle of Wight 20.11.47.
	G-AIWX	ex.	AX360	ca. 5.48, sold 5.50.
	G-AKVW	ex.	MG241	ca. 6.48, sold 8.50.
Avro 652A Anson 10	G-AIWV	ex.	NK668	ca. 8.47, ISOC.
	G-AIXU	ex.	NK823	dd.11.47, sold 6.49.
DH95 Flamingo	G-AFYH	c/n	95011	dd. 5.47, ISOC.
DH104 Dove 1	G-AKSS	c/n	04122	dd. 4.48, sold 2.49.

BRITISH AMERICAN

British American Air Services Ltd.

Bases: Heston, White Waltham and later Bovingdon Aerodromes.

British American Air Services started charter operations from Heston Aerodrome in April 1935 with a fleet of light aircraft. The directors of this company were Mr.D.S.Gibbs in North America,and Major M.Peto and Mr.L. H.Falk in London. Pleasure flying, passenger and freight charters were all undertaken, and the company specialised in the provision of passenger services to racemeetings. All operations were suspended, though, on the outbreak of war in 1939.

However, operations were resumed in January 1946 when the company started charter operations from White Waltham with one Dragon Rapide at the end of the month. Two more Rapides were acquired in February, and these aircraft entered service in March and April on passenger and freight charter work from White Waltham. These aircraft were again used for services to racemeetings, and also for several hops across the Channel to airfields on the Continent. However, the airline's mode of operations changed in July 1946 when British American Air Services took delivery of a Bristol Freighter leased from the Bristol Aircraft Company. Later, two other Bristol Freighters were bought, and these aircraft were used for a variety of tasks including the import of fresh fruit and vegetables from towns in France, Italy and Spain to markets in London. These aircraft were also used for other freight charters, including the carriage of racehorses, but losses incurred by these aircraft caused British American to sell these aircraft in the spring of 1947. By this time, the fleet of Rapides had also been retired, and in May 1947 a recession in the charter market led the company to further rationalise its operations, and all passenger charter work passed to Chartair who became an associate company of British American Air Services at this time.

This new partnership with Chartair led British American Air Services to purchase a fleet of all-cargo aircraft. The company had operated a Halton on behalf of the Maharajah Gaekwar of Baroda since March 1946, and although this aircraft was sold in April 1947, British American started its own Halifax freighter operations in July 1947. A second Halifax joined the fleet in September, and early in November the company moved its base to Bovingdon. From this new base, British American's first charter work consisted of a series of eighteen flights between November 8 and 20 carrying soft fruit from Valencia to Bovingdon. Throughout the next few months, this type of business proved to be a major source of income, and occasionally when work was heavy other aircraft had to be chartered to carry out the company's commitments. However, this work was interupted during December by a strike of people in France who made the packing cases for the fruit - one flight, though, was made from Nice to Bovingdon with six and a half tons of glace fruit. Work, early in 1948, built up rapidly and a third Halifax was added to the fleet early in March - charters at this time included two freight flights from Liverpool to Berlin, and two trips to Paris carrying three Bristol Hercules engines slung under each aircraft's belly. During April and May several flights were made to the Continent to collect fresh fruit and vegetables, and then on June 19 one of the Halifaxes left Bovingdon on lease to the Iraq Petroleum Company. This aircraft returned to Bovingdon on July 1 1948, having covered a total of nine thousand six hundred miles carrying general cargo, baggage, drilling machinery and mining equipment. On July 11 another Halifax carried five tons of penicillin and some valuable paintings from Bovingdon to Madrid - the paintings being carried in the aircraft's pannier, while the penicillin was stored inside the fuselage. Fruit flights, though, still continued, and on July 12 and 13 two Halifaxes brought apricots from Carmoli, in Spain, to Bovingdon. On their outward flights these aircraft would often position empty to collect this fruit, but on July 13 and 14 two flights to Paris carried general merchandise outbound, while on the return flights cargoes of greengages were carried. During August British American's Halifaxes imported many tons of bilberries from Southern France, and grapes from Spain, but at the end of the month British American's two aircraft both started work on the Northern Ireland milk airlift.

These two aircraft started to carry milk from Nutts Corner, Belfast, to Blackpool and Liverpool on August 31 1948. At this time, British American Air Services only had two Halifaxes in service and, therefore, during September and October 1948 no other charter work was undertaken. These aircraft returned to Bovingdon in the middle of October, and Airtech then converted them for use on the Berlin Airlift. Airtech was an associate company of British American Air Services, and although it was based at Thame, it set up a maintenance facility at Bovingdon to look after British American Air Services' aircraft. On December 11 1948 the two Halifaxes started flying supplies into Berlin from Hamburg, but one aircraft soon returned to Thame for conversion into tanker configuration. This aircraft returned to the airlift to start flying fuel oil into Berlin from Schleswigland on January 22 1949. The other freighter was then withdrawn from the airlift, and also converted to tanker configuration. One aircraft, though, was lost in an accident at Schleswigland on February 11 1949, but was soon replaced by another tanker Halifax. British American Air Services continued to fly fuel oil into Berlin from Schleswigland with two Halifaxes until being withdrawn from the airlift on July 12 1949, having flown a total of six hundred and sixty one sorties, of which only ninety seven had been freighters sorties, the remainder being tanker flights. The two tankers then returned to Thame for overhaul by Airtech, but the relatively high cost of converting them back for civilian freighter operations caused British American to only place one aircraft back into commercial service.

This surviving aircraft (G-AIAR) re-entered service with British American late in August 1949, and one of its first charters was a flights from Bovingdon to Milan and return on September 3. This charter was flown on behalf of Lep Air Services,and outbound the aircraft carried four tons of cargo, while on its return a cargo of six and a half tons of cloth was carried from Milan to Bovingdon. Flights carrying textiles were to provide British American with much of its revenue over the next few months - on September 27 and 28 two more textile flights were operated from Milan to Bovingdon. During the following week, a five and a half ton load of Russian furs was flown to Paris, and then the aircraft positioned to Milan to collect a further load of textiles.

In October 1949 Lep Air Services opened a new office at Southend Airport at the request of Southend Corporation, and on October 5 British American flew a load of over seven tons of textiles into Southend from Milan aboard its Halifax. The arrival of this aircraft marked the arrival of the largest load ever to land at Southend Airport, and many more flights were made to Southend by British American Air Services. On October 17 the Halifax was chartered by British European Airways to carry a consignment of radios to Greece, and once again on its return flight the Halifax stopped off at Milan to collect more textiles. During October and November, numerous textile flights were made from both Lille and Milan to Bovingdon and Southend on behalf of Lep, and very little other work was undertaken by the company. 1949, therefore, drew to a close with the airline operating almost solely between Milan, Lille and Southend carrying textiles. 1950 dawned, though, with a decrease in these flights and British American's surviving Halifax was forced to look elsewhere for ad-hoc freight charter work.

In February and March 1950 six flights were made to Stockholm carrying processed coffee, and in March several flights were made between Milan and Bovingdon carrying cotton yarn. In April several fruit and vegetable flights were made - these included flights from Amsterdam to Bovingdon carrying six tons of cucumbers on each trip. Early in May, six tons of chemicals were airlifted from Bremen to Bovingdon, and during June, July and August several flights were made between Milan and London carrying chemicals. However, work was becoming difficult to find and by the end of August 1950 the airline had stopped flying.

Among its last flights, British American's Halifax flew seven and a half tons of newspapers from Liverpool to Dublin on August 12, and two loads of silver Bullion weighting seven tons apiece were carried from London to Stockholm on August 11 and 22. Shortly after the return of the Halifax from Stockholm after the second silver bullion trip, the aircraft was withdrawn from service. The company had decided that there was insufficient business to keep this aircraft in service, and when its certificate of airworthiness expired, British American decided not to renew it, and thus British American Air Services ceased operations.

Fleet:

Fokker F.22	G-AFXR	c/n	5360	dd. 8.39,	sold 10.39.
DH85 Leopard Moth	G-ACUK	c/n	7080	dd. 4.35,	sold 2.37.
DH86	G-ADMY	c/n	2327	rd. 7.38,	sold 11.40.
DH89 Dragon Rapide	G-ADFX	c/n	6290	dd. 7.35,	sold 2.37.
DH89A Dragon Rapide	G-AGWC	c/n	6916	dd.12.45,	sold 4.47.
	G-AGZJ	c/n	6936	dd. 2.46,	sold 6.47.
	G-AGZK	c/n	6937	dd. 2.46,	sold 6.47.
Bristol 170 Freighter Mk.11A	G-AHJC	c/n	12735	lsd. 7.46,	ret. 9.46, leased from Bristols.
Bristol 170 Freighter Mk.11	G-AHJF	c/n	12738	dd. 8.46,	sold 3.47.
	G-AHJG	c/n	12739	dd. 9.46,	sold 4.47.
Handley Page 61 Halifax 8	G-AIAR	c/n	1388	dd. 2.49,	ISOC.
	G-AJPJ	c/n	1336	dd. 7.47,	sold 7.48.
	G-AKAD	c/n	1356	dd. 3.48,	written off Rennes 17.5.48.
	G-AKBB	c/n	1321	dd. 4.48,	written off Schleswigland 11.2.49.
	G-AKGN	c/n	1395	dd. 9.47,	written off Thame 17.12.49.
Handley Page 70 Halton	G-AGZP	c/n	1398	dd. 3.46,	sold 4.47.
Avro 652A Anson 1	G-AHUD			dd. 6.46,	sold 1.48, c/n quoted as RY/LW/1943.
Douglas C47A Dakota	G-AJRX	c/n	12209	dd. 2.47,	sold 8.47, never entered service.

BRITISH AND CONTINENTAL

British and Continental Airways Ltd. Base: Southend Airport.
British and Continental Air Charter and Freight Ltd.

British and Continental Airways was founded in 1946, and initially operated a Proctor 2 on light passenger and freight charter work from Southend and Gravesend. This aircraft also spent much of its time pleasure flying, and the company also purchased the remains of the Vega Gull G-AEXV which had crashed at Elmdon Airport in May 1946. By the spring of 1947, both aircraft were in service flying charters from Southend Airport, and on April 2 1947 another company named British and Continental Air Charter and Freight was registered as an airline by Mr.H.L.A.Powell and Mr.A.P.Claxton. This company had purchased a former Royal Air Force Avro Anson, and on May 10 1947 this aircraft entered service when it flew a freight charter from Southend to the Cape. Over the next eighteen months, this aircraft flew many long distance freight charters to the Middle East and Africa, but by the end of 1948 this company had ceased all air transport operations.

The name of British and Continental Air Charter and Freight did not disappear immediately, though, as the company took up residence in Airways House, Leigh-on-Sea, as travel agents. This activity, too, came to an end early in the 1950s, and the airline's name vanished forever.

Fleet:

Percival Vega Gull	G-AEXV	c/n	K.57	dd. .46,	ISOC.
Percival Proctor 2A	G-AIEH	c/n	H.66	dd.11.46,	ISOC.
Avro 652A Anson	G-AJSD	ex.	DJ165	dd. 5.47,	ISOC.

BRITISH CALEDONIAN

British Caledonian Airways Ltd. Base: London (Gatwick) Airport.
Caledonian // BUA Ltd.
Caledonian Airways Ltd.
British United Airways Ltd.

British Caledonian Airways is one of Europe's major airlines flying a network of both scheduled and charter services to many parts of the world. The scheduled network now extends to most major cities in East, West, Central and North Africa, South America and many European destinations including Amsterdam, Brussels, Genoa, Las Palmas, Lisbon, Paris and Rotterdam. In addition the airline also flies a network of domestic scheduled services, and there are also scheduled cargo services to several points in Africa.

British Caledonian's history can be traced back to 1970 when, in that year, the British and Commonwealth Shipping Company had been looking around for a buyer for its subsidiary, British United Airways. Talks were held with the two State airlines, but it was decided that it would be best to set up a 'second force' independent airline to compete with the Corporations on many of their routes. At a meeting on October 2[illegible] 1970 it was decided between Caledonian Airways and the owners of British United Airways that the former would take over British United on November 30. Under this agreement, Caledonian purchased the whole of the ordinary share capital of British United at a price of six million nine hundred thousand pounds. Under this new company - now named Caledonian // BUA - the scheduled services would be operated by British United Airways, and charter and inclusive tour services by Caledonian Airways (Prestwick) Ltd. A total of thirty one aircraft comprising VC10s, Boeing 707s and BAC 1-11s were operated by Caledonian // BUA, and some four thousand four hundred people employed including 450 flight deck personnel, 800 stewards and stewardesses, 1,000 engineering staff and 700 traffic staff. Mr.Adam Thomson became Chairman and Managing Director, while Mr.M.A.Guinane and Mr.John de la Haye became joint deputy Managing Directors of Caledonian // BUA.

Caledonian // BUA's route network was extensive, and when setting up this second force airline, it had been agreed that certain routes would be transferred from B.O.A.C. to Caledonian // BUA, and on March 31 1971 the airline took over B.O.A.C.'s services from London to Lagos, Kano and Accra. Then, on July 1 1971 the airline opened a Gatwick to Tripoli scheduled service. This was flown twice weekly with VC10s, and again was a route formerly operated by B.O.A.C. On November 1 1971 the airline, now known as British Caledonian Airways, started a service from Gatwick to Le Bourget, flown in competition with British European Airways and Air France.

In February 1972, British Caledonian was awarded its North Atlantic scheduled service licences. These were valid for fifteen years and were London - Birmingham - Manchester - Prestwick - New York, and London - Chicago - Los Angeles. Already the airline was operating a large number of charter flights across the North Atlantic from Gatwick, Manchester and Prestwick to the United States and Canada, but these affinity group charters tended to be seasonal, and it was hoped that the stability of operating scheduled services would help to more fully utilise the long range aircraft during the winter months. Transatlantic charters were also often flown from other airports - on October 7 1971 British Caledonian's flagship (G-AVKA) positioned from Gatwick to Rhoose Airport, Cardiff, to operate the first ever transatlantic charter from Cardiff. The aircraft was delayed at Cardiff by fog, but eventually left during the afternoon for New York, with a refuelling stop en-route at

Keflavik. The return service was flown by G-AVTW on October 24, arriving at Rhoose in the early hours of the morning direct from Kennedy Airport, New York. This aircraft was the first ever 707 to operate into Rhoose, and several other trans-Atlantic charters have been flown by British Caledonian from Wales' major airport since then. Meanwhile, the BAC 1-11s were also often used on charter work and many aircraft were used on inclusive tour charters from Bournemouth, Bristol, Cardiff, Gatwick, Glasgow and Manchester. However, this market was fast becoming too competitive and gradually British Caledonian reduced the number of aircraft allocated to this work, until by 1974 only one BAC 1-11 was used on inclusive tour charters. One of the airline's subsidiaries is Blue Sky Holiday Group which arranges package tour holidays to the Mediterranean and other resorts on both scheduled and charter aircraft.

On November 1 1972 BAC 1-11 G-AWYT inaugurated an Edinburgh - Newcastle - Copenhagen scheduled service - a Glasgow - Newcastle - Amsterdam service had been opened some years before by British United Airways, and this was also now operated by British Caledonian. Then, on January 1 1973 British Caledonian started to operate a London to Singapore charter service on behalf of Saber Air. Saber Air had previously operated these services with a DC8 leased from Overseas National, who had a minority shareholding in Saber Air and were responsible for its management. British Caledonian flew up to six services a week on this contract using its Boeing 707s. Further charters were flown from Singapore to Gatwick on behalf of the Chandris Shipping Line who used British Caledonian's 707s for a thrice weekly service carrying passengers from Singapore to London. These passengers had started their journey in Australia and had travelled from Australia to Singapore aboard one of Chandris' liners. These services continued until the end of the 1973 summer season.

In November 1972 the VC10s were withdrawn from the routes to South America which they had flown since their British United days. They were replaced on these services by 707s, and gradually the VC10s were replaced on all British Caledonian services. One of the VC10s was lost on January 28 1972 when it made a heavy landing at Gatwick after completing a training sortie - the aircraft was severely damaged structurally, and as the cost of rebuilding this aircraft would have been excessive, the aircraft was slowly dismantled for use as spares. During 1972 the 707s had been used to operate Bangladesh Biman's scheduled passenger services from London to Dacca until the arrival of their own aircraft to operate these services.

In April 1973 British Caledonian became the first independent airline to operate scheduled passenger services between London and New York. Amid a blaze of publicity and advertising on both sides of the Atlantic, on April 1 1973 came the big day when British Caledonian inaugurated its North Atlantic scheduled services. At Gatwick Airport a planeload of guests and passengers were piped aboard 'Flagship Bonnie Scotland' (now G-AWWD) for the inaugural flight which left Gatwick Airport at midday. Unfortunately, though, New York's weather was rather unfavourable and the aircraft diverted to Boston. British Caledonian also started the service from Gatwick to Los Angeles during April, and by August the airline's load factor on the Gatwick to New York service was fifty eight per cent, and on the service to Los Angeles sixty eight per cent. This encouraging start, though, was not to continue and at the ned of the first year of operations, an overall load factor of 41.4 per cent was achieved with a total of seventy thousand passengers carried. During 1974 it had been hoped to introduce services to Singapore via Bahrein, to Houston via Atlanta and Boston, and to Vancouver, but none of these services were started.

On November 1 1973 a service between Gatwick and Manchester was started. This route is flown twice daily by leased British Island Airways Heralds, and is timed to connect with British Caledonian's international scheduled services radiating from Gatwick. The airline also now uses British Island Airways Heralds for its Silver Arrow services between Gatwick and Le Touquet, which were previously flown with the 1-11s. On March 20 1974 British Caledonian became the first British airline to operate regular flights to Charles de Gaulle Airport, Paris. On this date the four daily Gatwick to Le Bourget flights were all transferred to Charles de Gaulle, and on April 1 British European Airways also started services from this new airport. Then, on June 4 1974 a Gatwick to Brussels service was opened, and using 1-11s this service is flown up to nineteen times weekly in each direction.

Although 1974 saw new services to the Continent being opened, British Caledonian was losing money on many of its services, especially those across the North Atlantic, and during September the airline announced serious cutbacks, both in staff numbers and services flown. Large numbers of employees were made redundant, including members of the management, and on October 31 1974 many services were axed. All North Atlantic scheduled services came to an end, the Gatwick to Belfast Interjet service passed to British Midland, the Southampton to Glasgow 1-11 service was taken over by British Airways' Viscounts, and the Edinburgh - Newcastle - Copenhagen service was closed down. In addition, many of the scheduled services to the Mediterranean came to an end including the routes from Gatwick to Gibraltar and Tunis. This reduction in services led to many 1-11s becoming redundant, and in order to combat this, many of these aircraft were put back into inclusive tour charter operation. Some aircraft, though, were placed into temporary storage until British Caledonian was able to arrange leases on these aircraft to many foreign airlines including Air Malta, Austral and Austrian Airlines. However, the Boeing 707 fleet has been increased to cope with the increased demand from advanced booking charters, with departures from the United Kingdom to airports in the United States, Canada and the Far East.

On November 1 1974 a weekly Boeing 707 service to Dakar, Senegal was started - this service was an extension of the existing service from Gatwick, via Casablanca and Freetown, to Monrovia. This service was followed on Thursday May 1 1975 by the opening of a service from Gatwick to Kinshasa via Lagos. In December 1975 British Caledonian announced a plan for cheap winter flights from Gatwick to Glasgow and Edinburgh - providing the passenger travels at the weekend and does not book his seat until the Friday before the flight, then he only pays twenty five pounds instead of the normal return air fare of forty four pounds. This system is similar to an earlier system adopted by British Caledonian during the winter of 1972 which was named 'Moonjet'. During 1975 British Caledonian suffered from competition by the B.E.A. London to Glasgow Trident shuttle service, and the airline applied to the Civil Aviation Authority for a fare differential on its London to Glasgow service in order to combat this competition.

The British Caledonian group also owns several major tour operators including Blue Sky Holidays and Golden Lion Holidays. In addition, a chain of ten hotels are owned in Spain, and British Caledonian's aircraft are used to carry many holidaymakers to Spain and other Mediterranean countries for their summer sunshine. On March 22 1976 British Caledonian expects to open a twice weekly 1-11 service from Gatwick to Algiers, while a twice weekly scheduled service from Gatwick to Tunis was expected to restart on March 24 1976. Looking further into the future, British Caledonian hopes to be able to restart trans-Atlantic scheduled services again - the airline is particularly interested in opening scheduled services to Atlanta and Houston. These two large American cities at present have no direct scheduled services to Europe, and are thus routes on which British Caledonian would have no competition.

Fleet:

BAC 1-11 201AC	G-ASJC	c/n	007	ISOF		'City of Glasgow'.
	G-ASJD	c/n	008	ISOF	sold 10.71,	'City of Edinburgh'.
	G-ASJE	c/n	009	ISOF		'City of Dundee'.
	G-ASJF	c/n	010	ISOF		'Burgh of Fort William'.
	G-ASJG	c/n	011	ISOF		'Burgh of Paisley'.
	G-ASJH	c/n	012	ISOF		'Burgh of Hawick'.
	G-ASJI	c/n	013	ISOF		'Royal Burgh of Nairn'.
	G-ASTJ	c/n	085	ISOF		'Royal Burgh of Dunfermline'.
BAC 1-11 501EX	G-AWYR	c/n	174	ISOF		'Isle of Tiree'.
	G-AWYS	c/n	175	ISOF		'Isle of Bute'.
	G-AWYT	c/n	176	ISOF		'Isle of Barra'.
	G-AWYU	c/n	177	ISOF		'Isle of Colonsay'.
	G-AWYV	c/n	178	ISOF		'Isle of Harris'.

G-ALEF "Red Eagle" - the start of what was to become British Eagle.

A J Jackson

G-AJBV was one of Eagle's many Vikings.

Flight

A later Eagle colour scheme was adopted when the airline was bought out by the Cunard Steamship Company. G-AIVO patiently awaits passengers on May 10 1961.

Ian MacFarlane

One of Cunard Eagle's Viscounts, G-ARKH, seen here on the apron at Elmdon Airport on April 29 1961.

Ian MacFarlane

Cunard Eagle's DC-6's operated a weekly flight to Bergen from Newcastle during 1962, G-ARWJ appearing on this service on many occasions.

Ian MacFarlane

What must surely be the best remembered aircraft type of the British Eagle fleet, the Britannia carried many thousands of tourists to their Mediterranean sunspots.

Ian MacFarlane

G-ASPN British Eagle Douglas DC-4 — B N Stainer/A.P.N.

G-AMHJ British United Island Airways DC-3 — Ian MacFarlane

G-APWI British United Herald — Transportraits/John Goring

G-AOGO British United Heron — Ian MacFarlane

G-AMHJ British Island Airways DC-3 — via A B Eastwood

G-BBXI British Island Airways Herald — Austin J Brown

BAC 1-11 501EX	G-AXJK	c/n 191	ISOF		'Isle of Staffa'.
	G-AXJL	c/n 209	ISOF		'Isle of Mingolay'.
	G-AXJM	c/n 214	ISOF		'Isle of Islay'.
BAC 1-11 509EW	G-AWWX	c/n 184	ISOF	sold 10.75,	'Isle of Skye'.
	G-AWWY	c/n 185	ISOF	sold 10.75,	'Isle of Iona'.
	G-AWWZ	c/n 186	ISOF	sold 11.75,	'Isle of Eriksay'.
	G-AXYD	c/n 210	ISOF		'Isle of Arran'.
BAC 1-11 515FB	G-AZPE	c/n 208	lsd. 8.72,	ret. 10.72,	leased from BAC.
BAC 1-11 530FX	G-AYOP	c/n 233	dd. 3.73		'Isle of Hoye'.
	G-AZMF	c/n 240	dd. 3.72		'Isle of Raasay'.
BAC VC10 1109	G-ARTA	c/n 803	ISOF	wfu. 1.72,	'Loch Ness'.
BAC VC10 1103	G-ASIW	c/n 819	ISOF	sold 11.74,	'Loch Lomond'.
	G-ASIX	c/n 820	ISOF	sold 10.74,	'Loch Maree'.
	G-ATDJ	c/n 825	ISOF	sold 4.73,	'Loch Fyne'.
Boeing 707-399C	G-AVKA	c/n 19415	ISOF	sold 5.73,	'Flagship Bonnie Scotland'.
	G-AVTW	c/n 19767	ISOF	sold 4.73,	'County of Ayr'.
Boeing 707-349C	G-AWTK	c/n 18975	ISOF		'County of Angus' later 'Flagship Bonnie Scotland' later 'County of Renfrew'. Registered G-BDCN 5.75.
	G-AWWD	c/n 19355	ISOF		'County of Argyll' later 'County of Angus'.
	G-BAWP	c/n 19354	lsd. 4.73,	ret. 5.75,	'County of Inverness', leased from Aer Lingus.
Boeing 707-355C	G-AXRS	c/n 19664	ISOF		'County of Caithness'.
	G-AYEX	c/n 19417	ISOF	ret. 10.71,	'County of Perth'.
	G-AYEX	c/n 19417	lsd. 3.73		'County of Argyll'.
Boeing 707-365C	G-ATZC	c/n 19416	ISOF		'County of Stirling'.
Boeing 707-323C	G-AYZZ	c/n 20089	lsd. 6.71,	ret. 9.71,	'County of Renfrew', leased from American Airlines.
	G-AYZZ	c/n 20089	lsd. 6.72,	ret. 12.73,	'County of leased from American Airlines.
Boeing 707-373C	G-AYSI	c/n 18707	lsd. 4.73		'County of Sutherland'.
Boeing 707-340C	G-AZPW	c/n 20275	lsd. 3.72,	ret. 9.72,	'County of Perth', leased from P.I.A.
	G-AZRO	c/n 20488	lsd. 4.72,	ret. 10.73,	'County of Lanark', leased from P.I.A.
Boeing 707-324C	G-AZJM	c/n 18886	dd. 3.73		'County of Ayr'.
Boeing 707-351B	G-BCLZ	c/n 18710	lsd. 9.74,	ret. 9.75,	'County of Lanark'.
Boeing 707-338C	G-BCAL	c/n 19297	dd. 5.75		'County of Lanark' later 'Flagship Bonnie Scotland'.
	G-BDEA	c/n 19296	dd. 5.75		'County of Inverness'.
	G-BDKE	c/n 19623	lsd.10.75,	ret. 12.75	leased from Qantas.
	G-BDLM	c/n 19629	dd. 1.76		'County of
Piper PA27 Aztec 250D	G-BBNN	c/n 27-4172	dd. 2.75.		

BRITISH EAGLE

British Eagle International Airlines Ltd.
Cunard Eagle Airways Ltd.
Eagle Airways Ltd.
Eagle Aviation Ltd.
Eagle Aircraft Services Ltd.
Eagle Airways (Bermuda) Ltd.
British Eagle (Liverpool) Ltd.
First Air Trading Ltd.
Air Liaison Ltd.

Bases: Aldermaston, then Luton, later Blackbushe, later Heathrow and Liverpool Airports.

Many will remember British Eagle as an example of one of Europe's finest ever independent airlines. From its birth in the spring of 1948 until its closure in the winter of 1968, British Eagle, and its predecessors, was continually in the public's eye fighting battles with the State airlines and licensing authorities. The scale and variety of Eagle's operations were such that rarely at any time during its twenty years of existence was the airline not involved in anything interesting, new or unique.

On April 14 1948 Mr.Harold Bamberg registered Eagle Aviation as an air charter company with a capital of one hundred pounds. Eagle's twenty five year old founder then purchased a fleet of Halifax freighters, and on August 26 1948 one of these aircraft started flying sorties on the Berlin Airlift from Wunsdorf. At the height of the airlift Eagle had three Halifaxes operating into Berlin, and on November 20 1948 these aircraft moved their base from Wunsdorf to Hamburg. Operations continued from here into 1949, and when British American Air Services was withdrawn from the airlift in July 1949, Eagle leased one of their Halifaxes to replace one of its own aircraft which had been withdrawn from service in June. Eagle Aviation was also, though, withdrawn from the airlift on August 15, and the fleet of Halifaxes was then flown back to the United Kingdom. Eagle set up its headquarters at Aldermaston Airport, and later in August the company bought out many of the aircraft previously operated by Blue-Line Airways at Tollerton. Before the airlift, Mr.Bamberg had also bought out Air Freight, and from this company Eagle inherited a fleet of Halifax aircraft which Bamberg was then able to use on the airlift.

At the end of the Berlin Airlift, therefore, Eagle Aviation had a varied fleet of aircraft available for charter - two of the airlift Halifaxes were kept serviceable for worldwide ad-hoc freight charter work, while the Blue-Line Anson, although initially advertised for sale, was later chartered to Transair. A variety of Auster and Proctor light aircraft were also available for air taxi work and pleasure flying. The Halifaxes, which had flown almost two and a half thousand hours on the airlift, found the post airlift period comparatively quiet although the occasional flight was made to the Middle and Far East carrying various cargoes including ship's machinery. The Halifaxes were now ageing, and it was soon realised that new equipment would be needed if Eagle was to pull through the post airlift depression in the air charter market. Thus, in October 1949, Mr.Bamberg announced that he had purchased a fleet of three Avro Yorks from B.O.A.C. This bold manoeuvre was to provide Eagle with the basis of its growth over the next three years. The aircraft could be fitted with either thirty or forty six seats for passenger charter work, or the seats could be removed altogether and the aircraft could be used as freighters. The first of these Yorks entered service with Eagle in November 1949, and early flights by these aircraft included two charters from Basle to Berlin on November 21 and 22 carrying cocoa. On the return trip from Berlin, the York stopped off at Amsterdam and collected a load of machinery which it then transported on to London. On November 26, one of the Yorks carried a load of radios to Athens, and then brought six tons of sweets from Istanbul back to London. As well as these freight charters, passenger charter work was also undertaken, and on December 7 a party of speedway riders was carried to Johannesburg in a flight time of fifty one hours. On its return trip the York collected a number of seamen in Athens, and then flew them back to the United Kingdom. The variety of Eagle's operations was further emphasised later in December when, on

December 20, a York flew a circus and its animals from Valencia to Liverpool for a Christmas circus taking place at Belle Vue, Manchester. The cargo carried included five lions, two monkeys, two ponies and five attendants to look after these animals during the flight.

The introduction of the York fleet had been smooth, and this resulted in the retirement of one of the Halifaxes in January 1950. This left just one of Eagle's faithful old Halifaxes in service which, during 1950, flew several ad-hoc freight charters. In addition to the Yorks, which were under the command of Captain J.W. Villa, Eagle had also bought a Dakota, and this aircraft entered service early in March 1950. One of its first charters was to bring a load of live lobsters, packed in cardboard boxes, from Stornoway to Aldermaston, where they were then transferred to a lorry and driven to Southampton. Early in 1950 Eagle Aviation was granted some B.E.A. Associate Agreements for the operation of scheduled passenger services. Mr.Harold Bamberg had always been an advocate of allowing private airline companies to compete with the State airlines by operating scheduled services. However, the routes awarded to Eagle were not exactly major trunk routes, and they were never operated for several reasons. The first route which Eagle had applied for was a London - Bristol - Cardiff service, but Eagle was only granted the London to Bristol sector, and without the extension to Cardiff the airline did not consider this route to be an economic or commercial proposition. The second route linked London with Plymouth and Penzance, and this route was not flown for two reasons. Firstly, another company had also been granted an associate agreement over what Eagle considered to be the most lucrative part of the route - the London to Penzance sector. And secondly, this route approval to the West Country was only for one route out of several applied for, and Eagle thought it uneconomic to fly just one scheduled service.

Although Eagle's efforts to set up a scheduled service network had initially been frustrated, the airline was later to prove its ability to operate these types of services. While the proposed scheduled service side of the business had come to nothing, the charter business was booming. Eagle was one of the few British charter companies to be operating comparatively modern four-engined long range equipment, and their aircraft were often chartered for flights to distant places. On March 18 1950 one of the Yorks took off from Bovingdon on the start of a three week time charter in South Africa under contract to the J.Arthur Rank Organisation The aircraft had been hired to carry a complete large screen television unit, consisting of studio equipment, a camera, transmitter and receiver. Ten technicians also accompanied this equipment, and while in South Africa the York also performed charter flights for other organisations. At the end of March another York flew forty six pilgrims from Teheran to Rome, and on April 1 1950 Eagle Aviation moved its operating base from Aldermaston to Luton Airport.

During April the company was busily engaged in carrying lettuces from Amsterdam to Bovingdon - a total of twenty four flights was made, with the first shipment arriving at Bovingdon on April 3 aboard one of the Yorks. In addition to the lettuces, Eagle also imported cucumbers from Holland - although the Yorks only brought in about six tons of lettuces on each flight, one York on April 7 landed at Bovingdon with over nine tons of cucumbers on board. The arrival of this aircraft constituted the largest ever load to have landed at Bovingdon, and for the remainder of the month Eagle's Yorks continued to ply between Amsterdam and Bovingdon with cargoes of lettuces, cucumbers and radishes for the London market of Covent Garden. Another heavy cargo was carried by Eagle during April when one of the Yorks transported a ship's propellor shaft, weighing nine tons and six hundredweight, from Thornaby to Amsterdam. The shaft was also the longest of its kind to have been carried by a British aircraft - its length was almost nineteen feet. Other charters carried out during April included the carriage of forty six passengers from Coventry to Amsterdam, forty one refrigerators to Dublin and one of the more unusual charters of the time - the transporting of sixteen tons of empty 'Pepsi-Cola' bottles from London to Cairo aboard two Yorks. This last charter was actually operated on May 1 1950, and at the end of May two more Yorks flew a load of mineral water preparation from London to Cairo. June saw the operation of further ad-hoc charter flights, and the Dakota, which had been equipped with thirty four passenger seats, was also busy flying passenger charters around the British Isles and Europe. On June 8 this aircraft flew a party of motor cycling enthusiasts from London to Ronaldsway for the Isle of Man TT Races. This Dakota left the United Kingdom for Malta in July 1950 to begin a seven month lease to Air Malta. While on lease to this airline, the Dakota flew scheduled passenger services from Luqa Airport, Malta, to Cairo, Catania and Rome. For the remainder of the 1950 season, therefore, Eagle was without any Dakota aircraft and the company concentrated on long range charter work with the Yorks.

During July one of the Yorks carried a seven ton ship's tail shaft from Amsterdam to Plaisance Mauritius. On the return flight from Mauritius, the York landed at Cairo to collect a Percival Proctor, on behalf of the Shell Petroleum Company, which was then brought back to Luton for repair by Percivals. At the end of the month another long range charter flight entailed the carriage of thirty seven seamen from Sao Paulo to Amsterdam. The aircraft departed from Luton on July 19, and after completing its work, it returned to the United Kingdom a week later. Another seamen's charter was operated on August 1 when a crew of twenty three were flown from Amsterdam to Abadan, and on arrival at Abadan the aircraft was turned round, and then left with a further twenty three seamen for the British Isles. En-route the York stopped off at Nicosia to collect a consignment of one and a half tons of gold residue. Other charters flown by Eagle's Yorks during August included the shipment of seven tons of cargo from London to Karachi on behalf of the Pakistan Government, and the carriage of an orchestra, its equipment and theatre artistes to Nuremburg. During September, one of the Yorks operated many pilgrim flights between Karachi and Jedda under contract to Crescent Air Transport. The last flight was operated on September 18, by which time a total of ten flights carrying four hundred and fifty passengers to Jedda had been made. September also witnessed the acquisition by Eagle of another company, Trent Valley Aviation, who had been operating passenger and freight scheduled and charter flights from Tollerton for several years. This company was bought out by Eagle on September 1 1950, and from Trent Valley, Eagle inherited a Dakota, a Rapide and a service from Tollerton to Jersey.

Trent Valley's Dakota moved south to Luton, and one of its first operations was to carry a party of Rolls Royce employees from Bovingdon to Paris on September 23. On October 26 it was announced that Eagle Aviation had been granted a B.E.A. Associate Agreement, valid for one year, for the operation of a scheduled passenger service from Tollerton to Jersey during 1951. On October 16 two Eagle Yorks left the United Kingdom for Singapore carrying ninety Royal Air Force personnel, while another York left for Iwakuni, near Tokyo, carrying a number of NAAFI personnel on October 14. During November the Yorks carried out a variety of charters - a B.O.A.C. sub-charter carrying penicillin to Cairo was undertaken on November 14, while on November 20 an aircraft left Bovingdon for Frankfurt and Rome to pick up parties of missionaries, who were finally deposited at Johannesburg. The company's Dakota carried a football team from London to Nantes on November 17, and a second York flight to Iwakuni carrying NAAFI personnel was made on December 2.

While the Yorks had been busy helping to establish Eagle Aviation as one of Europe's foremost charter airlines, the Halifax, too, had been actively engaged on freight charter work. During October 1950 this aircraft carried cargoes of chemicals, penicillin and silver bullion to Milan, and woollens to Munich and Dusseldorf. On October 25 the aircraft transported a five ton rudder from Luton to Singapore, and on November 15 1950 the Halifax left London bound for Singapore once again- this time under contract to B.O.A.C. However, ten days after leaving London the Halifax was lost in an accident at Calcutta, and after two years of service, Eagle's Halifax operations had now drawn to a close. The loss of this Halifax was not too serious to Eagle though, as in December 1950 Eagle purchased a fleet of three Avro Yorks from the Argentine Liquidation Commission at a price of nine thousand pounds per aircraft. Together with the purchase of these aircraft, Eagle also acquired a very large supply of spares for its Yorks, sufficient to keep all aircraft in the air for another three years.

An associated company of Eagle Aviation was Air Liaison Ltd. The Managing Director of this company was Mr.H.R.Bamberg, and his co-directors were Messrs J.D.Gazes and W.F.Lyons. Air Liaison acted as passenger and freight agents for all major airlines, and also shipping and general travel agents. The company occupied the same offices as Eagle Aviation in Clarges Street, London, and a secondary office was maintained in Birmingham. At the end of 1950 Eagle Aviation reopened the aero club at Tollerton with a small fleet of Austers The previous aero club there had been run by Trent Valley Aviation.

1950 had thus drawn to a close with business booming and Eagle operating one of the largest fleets of four engined transport aircraft in Europe. 1951 started with the usual series of long range charters, and increasingly now Eagle was being used by the Air Ministry for both freighting and trooping flights. On February 6 a York left Lyneham for Singapore with a cargo of Air Ministry aircraft parts. During January and February six flights were made to Gothenburg carrying machinery and electrical insulators outbound, and returning with reindeer meat - four of these flights to Sweden originated at Bovingdon, while the other two were operated from Liverpool. During March Eagle carried a consignment of freight under contract to B.O.A.C. to Hong Kong, brought a group of missionaries from Hong Kong back to London, and flew a party of forty five Rolls Royce employees to Paris for their Easter Holidays. The end of March saw Eagle in the Magistrates Court at Berkhamsted charged with infringements under the regulations that required an operator to have written loading instructions approved by the minister. The case was adjourned, though, and Eagle continued flying normally. Another of Eagle's marathon flights took place on April 5 1951 when a German cargo was flown from Amsterdam to Hong Kong. The aircraft then flew back via Delhi, where seats replaced the cargo interior, and nineteen seamen then clambered aboard for their flight back to London. While this charter was taking place, another of the Yorks carried more B.O.A.C. cargo to Hong Kong on April 11, and B.E.A. cargo was carried from Liverpool to Stockholm earlier in the month. During April and May, Eagle's Yorks were once again busily engaged in transporting vegetables from Amsterdam to Bovingdon, but these flights came to an end on May 22 when the Dutch Government stopped this operation. In June the scheduled service between Tollerton and Jersey was opened using a Dragon Rapide - the name of Trent Valley had been retained for this service because of the goodwill built up by this small airline in the Nottingham area.

August 1951 saw the operation by Eagle of the first regular contract flights on behalf of the Air Ministry. Prior to this, Eagle had been contracted to operate the occasional single flight, but during August Eagle Aviation was contracted to make four flights between the British Isles and Singapore alone. However, by the end of August seven flights, instead of the contracted four, had been operated, and more flights were undertaken during September. Eagle also converted two of its Yorks to carry stretcher cases, and Air Ministry work was now to become a very major part of Eagle's operations. September was, therefore, a busy month for Air Ministry work, but Eagle did manage to spare one York to carry three loads of pilgrims from Jedda to Kano between September 18 and 22. Eagle was contracted for further flights to Singapore during October, while the aircraft returning from Kano, in Nigeria, on October 1 collected some wild animals in West Africa and brought them back for clients in Amsterdam and London. The flights between Jedda and Kano proved to be particularly successful, and were flown on behalf of the Nigerian Pilgrim Society. During this operation Eagle made a total of six round trips - three flights from Kano to Jedda carrying pilgrims to Mecca, three flights from Jedda to Kano bringing the pilgrims back home, and six positioning flights. The success of this operation prompted Eagle to make arrangements early for the operation of similar flights in September 1952. The Air Ministry flights to Singapore continued throughout November, and two Yorks carried a thirteen ton cargo of cloth to Kano on behalf of the United Aircraft Company on November 23.

The normally quiet winter months were particularly busy ones for Eagle in 1951, with flights being made to many parts of the world - two flights were operated between London and Iwakuni early in December carrying service mail, and another York was chartered by B.E.A. to carry freight on the Nice - Rome - Malta - Athens - route early in the same month. On December 18 a War Office charter took servicemen from the United Kingdom to Nairobi, and NAAFI staff were flown from London to Nicosia on December 20 aboard Eagle Yorks. On December 9 one of the Yorks left Bovingdon on the start of a very long distance charter flight. The aircraft, with Captain C.W.Storm-Clark in command, originally left Bovingdon for Iwakuni, in Japan, with service mail. From Iwakuni the York flew to Hong Kong to collect a party of seamen, and take them to Rio Grande in Brazil. This party consisted of thirty Philippine seamen and one officer who were picked up at Hong Kong on Christmas Day and then flown to Porto Alegre, the nearest airport to Rio Grande. The aircraft then arrived at Luton early in January 1952 after covering a distance of over thirty one thousand miles. During December two Yorks flew cloth to Kano, Nigeria on behalf of the United Aircraft Company, and one of the Yorks then remained in Nigeria to undertake seventy hours of charter flying on behalf of United Aircraft. Another successful year had now drawn to a close, and Eagle now employed twelve pilots amongst its total staff of one hundred people. The share capital of Eagle was now fifty thousand pounds, and the directors of the company were Mr.Harold Bamberg, Mrs.Annis Bamberg, Mr.Charles Patterson and Mr.John H.Sauvage.

The New Year brought with it a number of Air Ministry flights - in January further flights were made to Singapore carrying service passengers, and on one flight stretcher cases were brought home. January also saw the award to Eagle of another relatively large contract by the Air Ministry. This contract called for the operation of about forty flights each year for two years, between the British Isles and Rhodesia. Although Yorks were initially used on these services, Eagle did open negotiations with the Ministry of Civil Aviation with a view to the purchase of up to six Short Solent flying boats for use on these flights. The sixty seat Solents would have been able to use the Victoria Falls area as their seaplane base in Rhodesia, but the negotiations for the purchase of these aircraft were not successful, and Eagle dropped this idea. The first service to Rhodesia left the United Kingdom on January 21 1952 flown by a York. During February an Eagle York brought twenty seven seamen from Delhi to London, Air Ministry flights were made to Singapore and Iwakuni, and six flights were made between the United Kingdom and the Egyptian Canal Zone on behalf of the Air Ministry. These flights to the Egyptian Canal Zone continued into March 1952, and during April and May many charter flights were undertaken by Eagle's Yorks. An eight ton load of metal castings was carried from Lyon to Manchester, a cargo of metal tubing was flown from Blackbushe to Catania, twenty nine seamen were taken to Malta from Luton, another ship's crew was brought from Amsterdam to Luton and further seamen were flown from Amsterdam to Oslo. These charter flights and the Air Ministry contract work were now Eagle's major sources of income.

The school and university holidays in July brought a large amount of student charters to Eagle - from July 5 1952 onwards Eagle's Yorks made regular trips from Luton and Bovingdon to Continental cities carrying parties of students. One of the Yorks was based in Copenhagen to operate flights to Zurich and Paris with Scandinavian students. When not required for these flights, the Yorks were used to carry Danish pigeons from Copenhagen to Stavanger and Amsterdam. On July 11 and 15 Eagle's Yorks carried the ninety members of the Turkish Olympic Games' team from Istanbul to Helsinki for the 1952 Olympic Games. The first of a new series of Air Ministry charter flights to Singapore began on August 7 1952 when a York left Blackbushe for Singapore with forty five servicemen on board. During August Eagle continued its student charter flights, and also flew the Turkish Olympic Games' team back to Istanbul.

The first of Eagle's ex-Royal Air Force Dakotas (G-AMPT) entered service early in September, and its first revenue earning flight was to carry twenty eight people from the United Kingdom to Lusaka on behalf of Lep, and a similar number of passengers were brought back to London on the return flight. This Dakota arrived back in this country on September 8, and the second ex-Royal Air Force Dakota entered service with Eagle on September 15 in a thirty two seat passenger configuration. Other charter work carried out by Eagle at this time included the carriage of fifty emigrant railway workers to Livingstone to join the Rhodesian Railways on September 14, and the airlift of four Hiller helicopters from Bourne Aerodrome, near Cambridge, to Khartoum for Pest Control. At the end of the month the Dakotas flew a ship's boiler to Marseilles, and carried twenty four passengers from Geneva to London.

On October 3 1952 one of Eagle's Yorks left Luton on the first stage of a charter flight that was to become the longest ever charter flight to be undertaken by a British operator. The first stage of this flight was to Karachi, where twenty cattle were collected and then flown to Belterra in Brazil. On the outward leg to Karachi, thirty Alsation dogs from London and Dusseldorf were carried on behalf of the Pakistan Government. The York, which was piloted by Captain T.Scott, was chartered by the Anglo Scottish Cattle Company, and from Karachi the aircraft flew via Aden, Khartoum, Kano, Dakar and Natal to Belterra. The York then returned to Karachi where a further twenty head of cattle were loaded aboard and then flown to Belterra, and then the aircraft positioned back to Luton after covering a total of thirty seven thousand miles on this charter. October also saw one of the Dakotas carrying a party of emigrants to Lusaka, and on October 27 twenty six seamen were flown from Stavanger to Amsterdam by Dakota. November was to be the last month of Eagle's York operations, and apart from three trooping flights to Nairobi, one of the aircraft flew forty two seamen from Amsterdam to Singapore. This aircraft left Amsterdam on November 9, but shortly after its return, Mr.Harold Bamberg announced that he had sold his fleet of Avro Yorks to Skyways. This sale took place on November 25 1952, and the price paid for these aircraft, and spares, totalled £160,000. In addition to taking over the Yorks, Skyways also took over the Eagle Air Ministry contract between England and Rhodesia which still had another fourteen months to run. In the year from July 1 1951 until June 30 1952 Eagle's Yorks had flown nearly a million miles, during which a total of 2,250 troops and 8,450 civilian passengers were carried. During the same period Eagle operated 673,000 ton miles of freight carrying and 6,664,750 passenger miles on private charter.

Thus Eagle had overnight become a Dakota operator, and all its long range trooping contracts came to an end. At the time it was stated that Mr.Bamberg had sold his Yorks as a protest at the Government's continued restraint on the operation of scheduled services by the private companies. However, Eagle settled down to its new role as a short haul operator, and in November bought out the Blackbushe based Aviation Servicing. This company was previously owned by Britavia,and with its purchase Eagle moved its operating and maintenance base to Blackbushe. During November the Dakotas were busy operating a variety of charter flights - half a ton of ship's spares were flown from Prestwick to Aden, twenty six seamen were flown from Liverpool to Antwerp and three tons of vaccine was flown from Palermo to Copenhagen. The last month of 1952 saw Eagle carrying newspapers from Manchester to Dublin, fifty Swiss agents of the Rootes Group from Zurich to Birmingham, cargo to Palermo,and a consignment of vaccine from Palermo to Copenhagen. By December,Eagle was also operating two Vikings which they had purchased from the defunct Crewsair- these aircraft were initially operated on the newspaper run between Manchester and Dublin which Eagle flew during November and December. 1952 closed with the operation of two inclusive tours by Dakotas on behalf of United States servicemen - one tour left Milan for a ten day tour of the Middle East on December 21, while a second group left Hamburg for a tour of Southern France and Spain on December 27.

1953 was to be a year of change for Eagle with the airline making headline news on several occasions. However, the year started undramatically with the operation of further newspaper flights between Manchester and Dublin, and more vaccine flights between Palermo and Copenhagen. Other points visited on both passenger and freight charters included Amsterdam, Kuwait and Zurich. One of Eagle's longer distance charters which was operated during the month was a flight from Liverpool to Dakar carrying a consignment of ship's spare parts. This flight took place on January 25 1953, and the 2,350 mile journey to Dakar was completed in twenty eight and a half hours. In January, Eagle announced that it was to lease its fleet of three Vikings to Airwork as it had been unable to fit the Viking into its pattern of operations. The two serviceable Dakotas had flown for two hundred hours on charter work during January, and during February these two aircraft carried out further charter work. A Bristol Hercules engine was flown out to Lisbon for Aquila Airways, while another aircraft engine, a Ghost, was flown to Paris for one of U.A.T.'s fleet of Comets.

Then, in March 1953 Eagle was granted a licence to operate a scheduled service from London to Belgrade via Munich. This licence was valid for seven years, and was Eagle's first chance to set up a network of scheduled services. In order to operate this and other new services Eagle acquired a fleet of Viking aircraft, but charter services continued during the spring and summer using both the Dakotas and Vikings. The Dakotas were booked by E.A.Gibson for a series of aerial cruises around the Mediterranean during the summer months. Over the Easter period early in April 1953, Eagle carried two parties of Catholics from Paris to the Holy Land, and another Dakota carried a party of United States servicemen on a tour of the Middle East. The first of three ex-B.E.A. Vikings entered service with Eagle early in April 1953, and one of the first flights by one of these aircraft was operated on April 6 when a group of twenty seven seamen were flown from Istanbul to Jarlesburg. In March 1953 one of Eagle's Dakotas moved to Berlin to start flying refugees out of the city to other German cities. This Dakota was later replaced on these services by one of the Vikings, and on May 18 1953 the Viking completed the one thousandth British sortie on this airlift. As the summer approached, Eagle's work increased and during the second week of May one of the Dakotas left for Baghdad under charter to Iraqi Airways for several weeks. Charter flights operated by Eagle during May included the carriage of a party from the British Grassland Society to Malmo, while a one and a half ton load of cinema equipment was flown to Dusseldorf later in the month. One of the Vikings flew a service to Nicosia, via Rome and Athens, on behalf of the NAAFI on May 18.

In May 1953 Eagle operated a proving flight between Blackbushe and Belgrade on behalf of the Press, and on June 6 one of the Vikings inaugurated this one weekly scheduled service. This event marked the opening of the first international scheduled passenger service to have been operated by a British independent airline since the last war. In order to operate this service and to develop others, Eagle founded a new company on July 1 1953 named Eagle Airways. In addition to this, the name of Eagle's maintenance company, Aviation Servicing, was changed to Eagle Aircraft Services on July 17 1953. Shortly after the inauguration of the service to Belgrade, another scheduled service, this time to Aalborg and Gothenburg, was opened by Eagle's Viking G-AJBO. Eagle Aircraft Services converted several of Eagle's Vikings to a comparatively comfortable high density layout. With thirty six reclinable seats facing aft, Eagle displayed the first aircraft to be converted (G-AHPO) late in July 1953. When used on trooping work these aircraft were known as 'Troopmasters', and when operated on civilian services they were named 'Mayfairs'.

At the end of 1953 Eagle had seven Vikings in service, and although only three hundred and fifty seven passengers were carried on its scheduled services during 1953, Eagle applied early in 1954 for a large number of other scheduled services. In February, Eagle applied to the Air Transport Advisory Council for permission to open scheduled services from Birmingham and Manchester to Cøpenhagen via Hamburg, Frankfurt via Brussels, Lisbon via Bordeaux and Madrid, Nice, Stockholm via Oslo, Tangier via Lyon and Barcelona, and Vienna via Lille and Munich. While the Air Transport Advisory Council considered Eagle's licence applications, Eagle continued to undertake much passenger and freight charter work. A strike of railwaymen on the Western Region of British Railways during May 1954 led to an emergency airlift of newspapers to South Wales and the West Country. Everynight for several weeks, Eagle's Dakotas and Vikings were used to carry newspapers from Northolt to Cardiff and Exeter until the end of the strike. In May the airline applied to add Strasbourg, Graz and Linz as intermediate traffic stops on their scheduled service to Belgrade. In addition to their scheduled services to Belgrade and Gothenburg during 1954, Eagle also operated a series of eight flights from Blackbushe to Belgrade carrying a load of twenty pigs on each flight. These charters were flown by Dakotas, and the first service left Blackbushe on June 15 1954.

On September 24 Eagle Aircraft Services showed to the Press and public, for the first time, their conversion of a Viking (G-AJPH) to freighter configuration. Eagle claimed that this aircraft had many advantages over its Dakotas as a freighter aeroplane. Apart from the better performance characteristics of the Viking, this aircraft had had all the passenger lighting removed, the centre bulkhead taken away and the fitting of a new freight door. In addition, a new large freight door (measuring 68in x 63in) was fitted, and Eagle claimed that with the incorporation of all these modifications that the main cabin area had been increased by twenty per cent. The first aircraft was 'blessed' by Mr.Peter Masefield at Northolt on September 24 1954, and the aircraft then entered service on Eagle's ad-hoc freight charter routes. The success of this conversion was confirmed when Airwork contracted several similar conversions of their own Vikings to Eagle Aircraft Services. Eagle's own Viking freighter came into its own only weeks after its entry into service - a dock strike in the United Kingdom had left goods stranded at ports all over the British Isles, and ports abroad were also unable to get their goods into this country. Eagle, therefore, joined with several other British companies and flew cargo to destinations all over Europe including Amsterdam, Gothenburg, Malmo and Paris. Throughout 1954 Eagle had continued to operate its scheduled services from Blackbushe to Belgrade and Gothenburg, and at the end of the year Eagle had ten Vikings in full service.

1955 was a year of expansion for the scheduled services network, and the company also ran a large number of trooping services to West Africa, the Mediterranean area and the Egyptian Canal Zone. In addition, Harold

Bamberg had also bought out the Sir Henry Lunn travel organisation, and all this company's inclusive tours were flown by Eagle's Vikings. During 1955 Eagle opened scheduled passenger services from Blackbushe to Luxembourg and Innsbruck, while from Manchester services were started to Munich and Vienna. These services were flown with Vikings, and also during the summer months Eagle operated inclusive tour services from Blackbushe to Corsica, Klagenfurt, Minorca, Nice, Palma, Pisa, Turin and Valencia. The Vikings made several trooping runs during the year from Blackbushe to Cyprus, Malta, Libya and Egypt on behalf of the Air Ministry, and by the end of the year the Viking fleet had increased to fifteen aircraft.

Eagle's 1956 inclusive tour programme was flown on behalf of several tour companies including Sir Henry Lunn, Wenger Air Tours and the Co-operative Travel Service. From Blackbushe the Vikings flew tourists to many resorts including Basle, Biarritz, Corsica, Klagenfurt, La Baule, Lyon, Minorca, Nice, Palma, Perpignan, Pisa, Rijeka, Zaragoza and Turin. For the first year ever, Eagle also flew inclusive tour charters from provincial airports - from May 1956 onwards Eagle's Vikings were used on services between Manchester and Palma and Birmingham and Palma on behalf of Sir Henry Lunn. In June 1956 Eagle introduced a twice weekly Viking service between Blackbushe and Dinard which was operated during the summer months only. The 1956 inclusive tour season was, therefore, a busy one, but Eagle still had many aircraft available for charter work, and when the Hungarian Revolution took place in November 1956, Eagle Airways was the second airline to arrive in Austria to help bring the refugees to Britain.

The evacuation of these people started on November 16 1956, and Eagle's first aircraft on the airlift (G-AMGG) arrived back at Blackbushe the following day with a load of refugees. This first mercy flight had been operated free of charge by Eagle Airways, and subsequently Eagle's aircraft flew regularly between Blackbushe and Vienna and Linz. On the outward trip the Vikings would carry blankets, food, clothing, vaccines, infants' necessities and much medical equipment. Every flight was accompanied by a Red Cross escort, and on the return flight from Vienna or Linz the aircraft would carry as many refugees as possible. When this airlift ended on December 14 1956, some seven thousand people had been airlifted out of Austria by British airlines. Of this total Eagle had brought over twelve hundred refugees (more than any other British independent airline) on thirty six flights.

Eagle's scheduled and inclusive tour services continued during 1957, and with further expansion in view an order was placed with Vickers for two series 805 Viscounts, with an option on a further three aircraft. An associate company of Eagle Airways was First Air Trading, and several aircraft were registered to this company, but all their aircraft were in fact flown in Eagle's colours. Eagle's trooping services in 1957 were still being flown by Vikings, and Air Ministry contracts took these aircraft to many points in the Mediterranean and West Africa carrying servicemen and their families. Three aircraft were based in Nicosia and these aircraft flew trooping services to Aden, Baghdad, Benina, El Adem, Malta, Mombasa and Tripoli from Nicosia. Eagle Airways was still solely responsible for the scheduled network, while Eagle Aviation looked after the airline's trooping, charter and inclusive tour services. Eagle's first Viscount was delivered to Blackbushe on December 23 1957, and this aircraft entered service the following day when it flew Eagle's first scheduled Viscount flight from Blackbushe to Hamburg and return.

As a prelude to the coming events of 1958, Harold Bamberg registered a new airline company named Eagle Overseas Airways on December 11 1957. The purpose of this company was to 'engage in the research and development of air transport and ancillary and cognate services in all parts of the World, and to establish, manage and operate air transport services'. This was followed by the setting up of a Bermudan subsidiary company named Eagle Airways (Bermuda) Ltd., and one weekend in April 1958 many people gathered at Blackbushe to watch the departure of Eagle's first Viscount (now registered VR-BAX) from Blackbushe for the sunshine of the Caribbean. Eagle Airways (Bermuda) had received a licence to operate a scheduled service between Bermuda and New York, and on June 2 1958 Viscount VR-BAX opened this new scheduled service. Meanwhile, Eagle's second Viscount G-APDX was retained in the United Kingdom for use on Eagle's network of scheduled services. During the summer of 1958 this Viscount flew scheduled services from Blackbushe to Pisa, from Birmingham to Palma, and from Manchester to Brussels, Copenhagen and Hamburg. The Vikings were used on the scheduled services from Blackbushe to Dinard, Innsbruck, Jersey, Luxembourg and Ostend; and from Manchester to Ostend. In addition, the company arranged for a coach service to operate from Luxembourg to Basle, and further licences were held for scheduled services from Manchester to Innsbruck and Pisa, and from London, Manchester and Glasgow to Reykjavik.

Eagle's Bermudan enterprise was to become an historic part of the development of British independent air transport. At the time of the Viscount's departure to Bermuda, Mr.Bamberg had stated that this, he hoped, would be a way in which Eagle could break the shackles that restrained it in the British Isles at that time. Mr.Bamberg appointed Mr.M.A.Guinane as the chief executive of Eagle's Western hemisphere division. On the route from Bermuda to New York, Eagle had competition from three other major airlines - B.O.A.C. who flew Britannias and later 707s, Pan American who flew 707s and Douglas DC8s, and Eastern Airlines who flew Douglas DC7Bs and DC8s. However, even with this competition Eagle carried almost ten thousand passengers during 1958 on its daily service to New York. In May 1959 a service was opened linking Bermuda with Washington/Baltimore, and later in the year, services were also opened to Montreal and to Nassau. An extension of the Nassau service to Miami soon followed, and in order to operate this service Eagle formed another company named Eagle Airways (Bahamas) Ltd. No aircraft were ever registered to this company, though, and they used the aircraft of Eagle Airways (Bermuda) for the Miami - Nassau - Bermuda service.

1959 saw a reduction in Eagle's European scheduled services, particularly those from Manchester, and the second Viscount was now transferred to the Bermudan operation. From Manchester, Eagle used Vikings on a fortnightly service to Ostend, and also on a fortnightly service to Bergen - both these services were flown on Sundays. The occasional charter was still undertaken by the Vikings and on July 21 a passenger charter was flown from Liverpool to Hamburg by G-AJBN. Eagle had also purchased a fleet of second hand Herons early in 1958, and although most of the aircraft were soon resold, one Heron did appear on several charter flights. On Wednesday September 2 1959 G-APKT operated the first of several charters from Hatfield to Middleton St.George via Blackpool. Eagle was also now operating a fleet of Douglas DC6s - its first aircraft (G-APOM) was delivered to Eagle's Blackbushe base on August 8 1958 with only twenty seven hours in the log book. This aircraft was so active that, after twelve months of intense operations, a total of 3,804 hours had been clocked up - a record at that time for a British registered aircraft. This vast total was logged between October 7 1958 and October 6 1959, and beat the previous record held by B.O.A.C.'s Argonaut G-ALHD by over two hundred hours.

Eagle also took delivery of two other DC6s at this time, and this fleet of aircraft was used on a wide variety of services during 1959. They were used on Air Ministry trooping flights from Blackbushe to Aden, Nairobi and Nicosia, but these flights came to an end in September 1959 when the contracts expired. However, ad-hoc Air Ministry contracts still continued, and during 1959 the DC6s visited many airfields while engaged on this work including Adelaide, Aden, Bahrein, Bombay, Christmas Island, Gan, Hong Kong, Kingston, Malta, Nairobi, Negombo, New York, Nicosia and Teheran. Of the commercial charters carried out, many were flown from the United Kingdom to Basle, Bradley Field, Buffalo, Keflavik, Kuwait, New York, Paris and Zurich. Flights were also operated between Amsterdam and Teheran, and between Tel Aviv and New York. The DC6s proved very popular in Eagle service, and during 1959 the airline applied to operate 'VLF' services to North America and the Caribbean. While awaiting approval for these services, Eagle operated a number of VLF charter flights to Bermuda. These flights, flown by the DC6s, operated via New York and cut the normal return air fare from £218 to £124. The operation of these services seriously affected B.O.A.C.'s traffic on these routes, and in consequence the 1960 Licensing Act reorganised the operation of these services. During 1959 Eagle Airways (Bermuda) carried over fifteen thousand passengers on its Bermuda to New York service, and its share of the market increased from just over six per cent in 1958, to almost ten per cent in 1959.

On Monday March 21 1960 it was announced that the Cunard Steamship Company had acquired a controlling interest of about sixty per cent in the Eagle group of Companies. Cunard became the seventh major shipping line to link with a British independent airline, and the shipping line's reputation Stateside was such as to give the Eagle group the impetus needed to open further services on this side of the Atlantic. Coincidental with the

Eagle take-over by Cunard was the introduction of Eagle's first Britannia. The aircraft concerned was a type 318 acquired on a seven month lease from Cubana, and delivery to Eagle was made on April 6. The aircraft had been delivered to Filton at the end of March and after an eight day overhaul programme, Eagle accepted delivery of the aircraft equipped with one hundred and four passenger seats. Its first service was flown on April 6 when a group of servicemen and their families left London Airport for Christmas Island in the Pacific. During the summer months, this aircraft was used on similar services to those of the DC6 fleet, and the Britannia's main occupation was initially the operation of many ad-hoc Air Ministry trooping flights. By July 1960 the Eagle companies were trading as Cunard-Eagle and most aircraft soon acquired these titles - Mr.Harold Bamberg joined the boards of the Cunard Steamship Company, and the Cunard White Star Line.

When Blackbushe Airport closed early in 1960, the Eagle group moved its main operating base to Heathrow Airport, and all the scheduled and inclusive tour flights previously flown from Blackbushe, were now flown from Heathrow. During 1960, Cunard-Eagle again operated fortnightly services from Manchester to Bergen and Ostend with Vikings, and in October 1960 the Britannia opened a service between London ,Bermuda and Miami. During 1960, almost twenty thousand passengers were carried on the Bermuda to New York service, and Cunard-Eagle's share of the market increased to just over twelve per cent. During 1960 the company also introduced two fifty seater Viscount 700s which it had bought from Cubana, while the two series 805 Viscounts were sold.

During 1961 Cunard-Eagle (Bermuda) operated two daily services from Bermuda to New York at the height of the season, and the routes to Miami and Nassau from Bermuda were flown four times daily. The London - Bermuda - Nassau - Miami route continued to be flown weekly by Britannias. These aircraft were fitted out with ninety eight seats - fourteen first class, sixty six economy class and eighteen Skycoach seats. During the summer of 1961 Cunard-Eagle's Britannias also operated seventy five transatlantic one-way charters, while the DC6s operated a further forty. During the year, Cunard-Eagle's share of the Bermuda to New York run rose to fifteen per cent, and the airline also introduced several off-peak air fares in order to try to attract further business. While its Bermudan activities were being built up, Cunard-Eagle was not forgetting its home market in the British Isles, and in August 1961 a series of licence applications were heard by the Air Transport Licensing Board. These applications were for services from Gatwick, via Liverpool and the Isle of Man, to Belfast; from Gatwick, via Liverpool, to the Isle of Man; from Gatwick to Liverpool; from London to Edinburgh; from London to Glasgow (Renfrew); from London to Birmingham; and from London to Manchester. These applications caused a large number of airlines to lodge objections, including B.E.A. and Starways. Most of the applications were unsuccessful, although some of the routes were later approved and opened by Eagle. During 1961 Cunard-Eagle spent much of its time battling with the A.T.L.B. for permission to operate scheduled passenger services, in competition with B.O.A.C., over the route from London to New York. Surprisingly enough, Cunard-Eagle became the first independent to be granted just such a licence, and as a result two Conway powered Boeing 707s were ordered for 1962 delivery. Sadly, though, B.O.A.C. objected to another competitor on this prime North Atlantic route and the State corporation appealed against the licensing board's decision. Subsequently, B.O.A.C.'s objection was upheld, and Cunard-Eagle's licence was revoked before any flights had been made.

In July 1961 Cunard-Eagle applied for permission to operate its services to Ostend,Luxembourg and Innsbruck from both Heathrow and Gatwick. During the summer of 1961 Eagle operated two Viking flights a week on the London to Luxembourg run, one Viking weekly on a London - Ostend - Luxembourg service, three DC6 services weekly between London and Innsbruck and one DC6 service weekly on the route London - Luxembourg - Innnsbruck. Late in August, Cunard-Eagle signed a pooling agreement with Austrian Airlines for the London to Innsbruck services. The variation of its licences to allow it to operate the services to Austria from Gatwick could be expected, as Cunard-Eagle Britannias and DC6s were nightly visitors to Gatwick on Air Ministry trooping flights during 1960 and 1961. During the summer of 1961, Cunard-Eagle's services from Manchester linked this northern city with Bergen, Hamburg, Copenhagen, Ostend, Pisa and Rimini, and on many of these routes the airline used its Viscount 700s.

With the revokation of Cunard-Eagle's London to New York licence, the airline was to take delivery in March 1962 of the first of two redundant Boeing 707s. Although these aircraft would find temporary work on the London to Bermuda route, the aircraft would have to be Bermudan registered as Cunard-Eagle's Bermudan subsidiary was the company that held the licence for this route, the British company did not. In December 1961, though, Cunard Eagle, realising that there was a difficult time ahead, announced cutbacks in many of its services. The airline's two European based Viscounts (G-ARKH and G-ARKI) were put up for sale, and the summer services from Manchester to Bergen, Ostend, Pisa and Rimini would not be resumed the following summer as originally planned. The services to Hamburg and Copenhagen from Manchester had already closed in October 1961, and these routes, too, were not to be reopened. In making these cuts, Cunard-Eagle stated that the closure of these services was due to its inability to build up a network of viable scheduled services radiating from Manchester For several years the airline had been applying for more scheduled routes from Manchester, but the A.T.L.B. continued to turn down Cunard-Eagle's persistent applications. Eagle applied three times for a Manchester to London service alone, and each time the result of the hearing was an emphatic 'no'. Eventually, Eagle decided to give up its fight for these services, but over the next few years the airline was to have numerous battles with the authorities over its licence applications.

By early 1962 Cunard-Eagle had still not found a buyer for its two Boeing 707s, and so preparations were made to accept delivery of these aircraft. Cunard-Eagle's chief pilot, Captain H.Wilkins, nine other captains and seven flight engineers all left London for training with Boeing in North America in preparation for the entry into service of a British independent airline's first pure jet aircraft. Cunard-Eagle went ahead with its plans, and on February 27 took delivery of the first Boeing 707. This aircraft, operating under Bermudan registry, was put into use on an ad-hoc basis on the route between New York and Bermuda from March 27 onwards, supplementing the existing Viscount schedules. Then, on May 5 1962 Cunard-Eagle proudly inaugurated a twice weekly Boeing 707 service between London , Bermuda and Nassau. On board this inaugural flight, piloted by Captain Henderson, were several guests of Cunard-Eagle, while the airline itself was represented by Mr.Ben Russell, a director of the Cunard Steamship Company, and Mr.Harold Bamberg himself. Before the inauguration of this service, Cunard-Eagle were fortunate enough to obtain traffic rights into Jamaica, and plans were made for a once weekly service to Montego Bay as an extension of the London to Nassau service. However, these services were insufficient to keep two brand new Boeing 707s fully utilised, and although Cunard-Eagle had hoped to help in the formation of a national airline for Jamaica in August 1962, it was soon accepted that this precarious position could not go on for much longer.

In June 1962, the Cunard Steamship Company got 'cold feet' and lost confidence in its airline associate, and with B.O.A.C., Cunard formed a new company named B.O.A.C.-Cunard to operate its services. The formation of B.O.A.C.-Cunard took place on June 20 1962, with B.O.A.C. taking a seventy per cent shareholding, and Cunard taking up the remaining thirty per cent. This resulted in a row breaking out between Harold Bamberg and the Cunard Steamship board, but the days of Cunard-Eagle were numbered. Arrangements were then made for the transfer of Cunard-Eagle's Bermudan services to B.O.A.C.-Cunard. The second Boeing 707 was delivered direct to B.O.A.C.-Cunard on July 7 1962, and most of the Cunard-Eagle staff based in Bermuda and New York were also taken over by B.O.A.C.-Cunard. Cunard-Eagle Airways (Bermuda) continued to operate its services under its own name until September 30 1962 when all its services were formally transferred to B.O.A.C. The services from Bermuda were taken over by B.O.A.C.'s associate, Bahamas Airways, and the run from London to Bermuda and Nassau became a B.O.A.C.-Cunard service, flown by their Boeing 707s. Prior to this transfer, Cunard-Eagle had suspended its New York to Bermuda Viscount service on September 9, and with the closure of all its services in the Caribbean, Cunard Eagle became an all European based airline once more.

Although its Bermudan operations had come to a sad end, Cunard-Eagle's European services were now being developed once again. On May 20 1962 one of the airline's DC6s (G-APON) opened a Manchester - Newcastle - Bergen service which was flown weekly throughout the summer until September 23 1962. Another DC6 (G-ARZO) inaugurated the first service on a regular Air Ministry contract to Adelaide on June 8 1962, and this contract was retained by Eagle until its closure in 1968. 1962 had thus been a turbulent year for Cunard-Eagle -

G-APWA has flown with many airlines all over the world - British Midland's turn coming in 1966.

Ian MacFarlane

G-BAVX has not flown extensively for British Midland yet, but it did open the airline's new service between Elmdon and Heathrow on November 1 1975.

Tony Merton Jones

British Midland's Viscount G-BAPF leaps off St. Mawgan's lengthy runway for another pleasure flight at the 1974 St. Mawgan Air Day.

Tony Merton Jones

Derby Airways was the only British airline ever to operate the Miles Marathon, and three aircraft were used by Derby on its scheduled services until the end of the 1960 season.

P H Butler

One of British Midland's three Argonauts, G-ALHY, is pictured here on the apron at Liverpool Airport.

P H Butler

G-AYVE, a Boeing 707 of British Midland Airways, awaits its passengers at Prestwick while operating a trans-Atlantic charter flight for British Caledonian.

Gordon Bain

G-AIUK **Derby** Aviation Dragon Rapide A J Jackson

G-ALHY Derby Airways Argonaut Ian MacFarlane

G-AWXI British Midland Viscount 814 Ian MacFarlane

G-AKOV Wolverhampton Aviation Dragon Rapide A J Jackson

G-AWCV British Midland Viscount Ian MacFarlane

G-AROI British Westpoint Dove Transportraits/P R Keating

although it had seen the operation of its first jet service, it had also seen the termination of some five years of hard work in the Caribbean to develop scheduled services in competition with three of the world's largest airlines. However, Cunard-Eagle still had a large European scheduled and inclusive tour netweork to fall back on, and with a large number of ad-hoc Air Ministry trooping and freighting contracts still coming its way, Cunard-Eagle could still look forward to a rosy future.

On February 14 1963 Mr.Harold Bamberg repurchased a sixty per cent shareholding from the Cunard Steamship Company, and Eagle once again came under the control of Mr.Bamberg, and this effectively severed all links with the Cunard Steamship Company. This transaction completed, Bamberg then set about moulding his airline into a major European carrier, and during the summer of 1963 the airline was operating a larger European scheduled service network than that of any other British independent airline. Further expansion plans were announced late in August 1963 when Eagle was granted to operate scheduled services from London to Edinburgh and Glasgow. This was the first time that a British independent airline had been allowed to operate services in parallel to those of one of the State Corporations. On August 9 1963 Cunard-Eagle changed its name to British Eagle International Airlines in preparation for the introduction of the newly approved scheduled services. The first of these services to be opened was the London to Glasgow route - on October 22 Britannia G-AOVT operated a proving flight from Heathrow to Renfrew and return, with representatives from the Press and guests of British Eagle. The first revenue earning flight over this route took place on Sunday November 3 - British Eagle used Britannias on its services equipped with fourteen first class seats, and eighty seven tourist class seats. This service was flown six times weekly in each direction. The second of the services to be opened was the trunk route from Heathrow to Edinburgh, which was opened by Britannia G-AOVT on November 4 1963. British Eagle flew a daily weekday service over this route initially, and services were also opened from Heathrow to Belfast during November.

Further expansion of British Eagle's domestic network came on November 19 1963 when British Eagle signed an initial co-operation agreement with the Liverpool based airline of Starways. This was in effect a take-over of Starways by British Eagle, and from this Liverpool based airline, British Eagle inherited a number of important routes. These included the lucrative London to Liverpool service, and summer only services to Newquay from several departure points in the United Kingdom. Also operated by Starways were services from Liverpool to Cork, Glasgow and Edinburgh; from Blackpool to Glasgow, and from Hawarden to Heathrow. Starways also operated a large number of inclusive tour charter flights from Liverpool and Manchester with its fleet of Dakotas, DC4s and Viscounts. However, British Eagle's take-over only included the route network of Starways - the aircraft were retained by some of Starways' directors and later sold. One of Eagle's Viscounts appeared with Starways' titling, but on January 1 1964 British Eagle officially took over all Starways' services, and a new company named British Eagle (Liverpool) was formed to operate all Liverpool based services, and this division also later became responsible for most of British Eagle's Viscount maintenance.

On January 1 1964 British Eagle increased the frequency of the London to Belfast service from five to six services weekly. The London to Liverpool service was flown by Viscounts and Britannias, and during its first month of operations over this route, British Eagle carried over five thousand passengers. On April 1 1964 a second daily Britannia flight was introduced on the London to Liverpool route, and another thrice weekly Viscount service on May 12. For the summer of 1964, British Eagle introduced new holiday routes from London to Brindisi (flown once weekly) and from Manchester to Nice (flown twice weekly). The airline also introduced new scheduled services to Stuttgart from London, and to Rimini and La Baule from Manchester. British Eagle also started a thrice weekly scheduled service from Liverpool and Birmingham to Palma - flown at night by Viscounts this service operated on Tuesdays, Thursdays and Saturdays throughout the summer. During the summer, British Eagle's network of scheduled services consisted of seventeen routes out of London, and a further fifteen routes out of six provincial cities. The Starways network of services to Newquay was also reopened in May, but the service from Newcastle to Newquay was closed down. Inclusive tour charter work was very buoyant during 1964, and in order to undertake Starways' 1964 inclusive tour programme from Liverpool and Manchester, British Eagle bought a fleet of three DC4s. Other inclusive tour flights were operated from Heathrow, Birmingham and Manchester on behalf of several tour companies including Lunn-Poly and Everyman Holidays. In 1964 the remaining DC6s in the Eagle fleet were replaced by two Britannia 308s. British Eagle bought these aircraft from the liquidators of Transcontinental SA., and the first aircraft arrived at Heathrow on January 22. The two aircraft were converted to freighter configuration by the fitting of a special cargo door measuring six feet by seven feet four inches, similar to the doors fitted to the Royal Air Force Britannias. After conversion, the first aircraft entered service in April, replacing the DC6s on the Government freight contract from London to Adelaide.

During 1964 British Eagle continued its now traditional battle with the licensing authorities by once more applying for a London to Manchester scheduled service. Not unexpectedly the application was turned down by the A.T.L.B., but Eagle was allowed to increase the frequency of its services from London to Glasgow and Edinburgh. As from April 1 1964, British Eagle was allowed to raise the frequency of the London to Glasgow route from seven to twelve services weekly, and the London-Edinburgh route from seven to ten services weekly. Although these frequencies were still well below those requested, it was at least a step in the right direction for Eagle, but an application to increase the number of services to Belfast met no success. Later in the year, the airline once again applied for further frequency increases, but to no avail, and consequently British Eagle withdrew its London to Belfast and London to Edinburgh services on February 20 1965.

The next 'stab in the back' for British Eagle came when B.O.A.C.-Cunard applied to the A.T.L.B. in June 1964 for the revokation of a licence held by British Eagle International which allowed it to operate transatlantic services to Canada and the United States of America. This licence covered flights to Montreal and Toronto in Canada, and to Boston, Chicago, Detroit, Baltimore, New York, Philadelphia and Washington in the United States. Although this licence had never been used by British Eagle, it was at this time that with the acquisition of many suitable long range aircraft, the airline was considering opening a series of inclusive tour charter flights to North America and this licence would have been a way of starting these services. During 1965 the airline continued to build up its fleet of aircraft and the DC4s were sold and replaced with further Viscounts. Apart from the suspension of the London to Belfast, and London to Edinburgh services, all routes were flown during the summer of 1965.

While 1965 was a year of consolidation, 1966 was the year in which British Eagle returned to pure jet operations. Central African Airways (now Zambia Airways) and Kuwait Airways had both ordered fleets of BAC 1-11s which they no longer required. Thus, British Eagle leased these aircraft, and on April 22 the first aircraft arrived at Heathrow on delivery to the airline. After repaint, this aircraft was rolled out as G-ATTP 'Swift' on April 30 1966, and on May 2 1966 it became one of the first aircraft to land at the newly opened airport at Abbotsinch, Glasgow while engaged on crew training. British Eagle had hoped to introduce the 1-11 onto its London to Glasgow service on April 18, but this had not proved to be possible, but on May 9 1966 G-ATTP inaugurated British Eagle's first BAC 1-11 service when it flew a scheduled passenger service from London to Glasgow. The second 1-11 was delivered on May 27 1966, and this aircraft first appeared on the Glasgow service three days later on May 30. On May 28 British Eagle placed 1-11s onto the service from London to Tunis and Djerba, and the following day saw the operation of 1-11 services to Luxembourg and Stuttgart from London. On Tuesday May 31 the 1-11 appeared on the London to Dinard and La Baule service, and by July these aircraft were also flying scheduled services from London to Liverpool, Newquay, Palma, Perpignan, Pisa and Rimini. The 1-11s were also used on inclusive tour charter flights from Manchester and Heathrow to many Mediterranean resorts on behalf of the Travel Trust Holiday Group (which included Lunn-Poly and Everyman Holidays) and Global Holidays. The 1-11 proved to be ideal for British Eagle's services, and soon proved popular with passengers and crews alike.

Another 'new' aircraft entered service with British Eagle in 1966 - on July 4 a Dove started flying a feeder service from Dundee to Glasgow to connect with British Eagle's scheduled services to London. This aircraft operated two services each weekday in each direction, and was affectionately named 'Eaglet'. During 1966 British Eagle carried just over ninety four thousand passengers on its international scheduled services, and two hundred and ten thousand passengers on its domestic scheduled services at an overall load factor of just over forty seven per cent. At the end of 1966 a reorganisation of the Eagle group of companies was made - under this reorganisation a new holding company named Eagle International was formed. Its subsidiary companies included British Eagle International Airlines, British Eagle (Liverpool), Knightsbridge Air Terminal, Eagle Aircraft Services and Sky Chef. At the same time Mr.Bamberg repurchased the remaining forty per cent of the Cunard interest, and British Eagle became wholly owned by Mr.Bamberg and his associates. During 1966 British Eagle carried almost one million passengers on its scheduled and charter services, and on a turnover of fifteen and a half million pounds, the profit made was £585,000. This compared very favourably with the results obtained three years before - in the first year of British Eagle's operations in 1963 the airline carried only one hundred and fifty three thousand passengers, and there was an overall loss of £80,000.

During 1967 British Eagle made few changes - further 1-11s were acquired early in the year, and an order for two Boeing 707s was placed with Boeings. The aircraft were ordered for use on a network of transatlantic scheduled services proposed by British Eagle. During 1967 the airline applied to the A.T.L.B. for permission to operate services from London to New York, the Bahamas, Bermuda, Jamaica and Hong Kong, for introduction in the spring of 1968. An application was also made to operate an all-freight scheduled service between London and the eastern seaboard of the United States. During the summer, British Eagle operated all its normal scheduled services, and at the end of the season the company arranged leases for some of its 1-11s to other operators. One aircraft (G-ATPJ) went on lease to K.L.M. to operate their London to Rotterdam services, while two other aircraft (G-ATVH and G-ATPL) went on lease to Swissair and S.A.S. respectively. Two of the company's Viscounts were sold to Invicta Airways, and on November 1 1967 British Eagle withdrew its scheduled service which linked Cork with Liverpool. This route had been flown since the Starways take-over, but continuing losses eventually forced British Eagle to withdraw from this service, which had been flown in association with Cambrian Airways. During the winter of 1967, British Eagle started flying tourists from London to East Africa by Britannia and plans were laid during the winter months for the introduction of inclusive tour services to Bermuda and the Bahamas for the following summer. Over three hundred and twenty five thousand passengers were carried on British Eagle's scheduled services during 1967, while the aircraft flew an impressive total of fourteen million miles on non scheduled services.

On December 29 1967 the final chapter of British Eagle's history started when a Qantas Boeing 707 VH-EBD started crew training for British Eagle crews at Sydney Airport. On January 4 the aircraft was registered G-AVZZ, and on February 3 it arrived at Heathrow Airport in full British Eagle colours. The first of British Eagle's two brand new 707s was handed over to the airline at Seattle on December 22 1967, but the airline was still battling with the A.T.L.B. for its North Atlantic scheduled service licences. The aircraft (G-ATZD), therefore, remained in North America until British Eagle decided what to do with this aircraft. Then, on February 4 1968 this aircraft arrived at Shannon as VR-BCP, and left for Rome two days later on lease to Middle East Airlines. This aircraft was leased to Middle East Airlines pending a decision by the A.T.L.B. on British Eagle's North Atlantic licence applications. Meanwhile, British Eagle's former Qantas 707 was engaged on crew training sorties at Prestwick and Stansted during February.

Almost six years to the day after Cunard-Eagle's first Boeing 707 arrived in Bermuda, G-AVZZ operated a proving flight late in February 1968 from Heathrow to Nassau in preparation for the summer's inclusive tour flight programme to the Caribbean. A second Qantas Boeing 707 arrived at Heathrow on March 15 on delivery to British Eagle, and during the summer these two aircraft were used on inclusive tour charters from the United Kingdom to the Caribbean, and also on inclusive tour flights from London to Alicante, Gerona, Malaga, Palma, Rimini, Rome, Santiago, Split, Tenerife and Venice, and from Manchester to Palma. The 1-11s flew inclusive tour charter flights from Heathrow to Alghero, Dinard, Djerba, Ibiza, Istanbul, Gerona, La Baule, Luxembourg, Mahon, Palma, Pisa, Rimini, Stuttgart and Tunis; from Liverpool to Ibiza, Palma and Rimini; from Birmingham to Ibiza and Palma; and from Manchester to Alicante, Gerona, Ibiza, Malaga, Palma, Rimini, Tenerife and Venice.

However, all was not rosy, and in March 1968 British Eagle's trooping and Australian migrant contracts came to an end. These two activities had been a major source of income for British Eagle, and their loss was followed by a bad summer. A rather less important loss was the closure of the Dove feeder service between Dundee and Glasgow in the spring of 1968 which had run for almost two years, but was suspended when Autair introduced a London to Dundee Herald service. During the year the company had suffered cancellations in inclusive tour holiday bookings of over one million pounds. Bamberg, though, pressed on with his expansion plans and Eagle International (Bahamas) applied to operate scheduled services from Nassau to Antigua, Barbados, Bermuda, New York, Toronto and Washington. At the same time Eagle International (Bermuda) applied to operate similar services from Bermuda to New York, Toronto and Washington. Eagle was never to operate scheduled services from Bermuda or Nassau again, though, and the first sign of an impending collapse of British Eagle came early in October 1968.

In October, four hundred British Eagle employees suddenly found themselves without jobs, and the base at Liverpool was closed. During 1968 British Eagle had been financed by two banks, Hambros and Kleinwort Benson, and new financial arrangements were finalised for 1969 on October 30 1968. However, within days of these arrangements, British Eagle suddenly had its inclusive tour licence to the Bahamas and Bermuda revoked. This was followed by both banks withdrawing their support from Eagle, and the collapse had started. Although the services to the Caribbean accounted for only three per cent of the airline's total flying proposed for 1969, and that Eagle had its scheduled services from the Bahamas to New York and Canada approved early in November, the banks were still not convinced that British Eagle's future was assured. Unfortunately, most of the finance was needed to support British Eagle through the coming winter, and with no other finance available the collapse became inevitable.

Early in the evening on Wednesday November 6 1968 the British Eagle board announced that at midnight all flying by the airline would cease. This stunning announcement was followed by the return to Heathrow and Liverpool of all British Eagle's aircraft. The last aircraft to touch down at Heathrow was a Britannia freighter G-AOVM from Israel with a cargo of oranges, and this aircraft landed on November 7.

At the time of the collapse, creditors were owed just over five and a half million pounds, and although the airline's assets (at book value) were worth over five million pounds, the stock when sold did not raise anything like this figure. The list of creditors included Rolls Royce (owed £630,000), Esso (£300,000), Elliots (£36,000), Dunlop (£30,000), and Vickers (£10,000). The airline's board of directors issued a statement after the closure which attributed the airline's collapse to 'devaluation, the £50 travel allowance and the present economic situation which gave rise to a severe recession in holiday travel' - the company had suffered cancellations in inclusive tour holidays of £1,075,000 for the 1968 summer season.

During its twenty years of flying, Eagle had never had an industrial dispute or a strike amongst its employees, and at the time of the collapse some two thousand three hundred people were employed by the airline. Many members of both the management and the workers tried valiantly to keep the Eagle flying, and proposals were put forward for the formation of a new airline flying a smaller number of Britannias and 1-11s. Sadly, though, none of the plans ever came to fruition, although companies like Fairflight Charters and Pheonix Airlines were established by former Eagle employees early in the New Year, and many staff went to work for the newly formed Donaldson International. Truly, this was a sad ending for an airline that had played a very major role in re-establishing the private sector of British air transport post war.

Fleet:

Handley Page 61 Halifax 8	G-AIAP	c/n	1354	dd. 2.49, written off Calcutta 25.11.50.
	G-AIAR	c/n	1388	lsd. 7.49, ret. 8.49, leased from B.A.A.S.
	G-AJBL	c/n	1349	dd. 7.48, wfu. 7.49.
	G-AJCG	c/n	1390	dd. 8.48, wfu. 6.49.
	G-AKBR	c/n	1391	dd. 8.49, wfu. 9.49.
	G-ALEF	c/n	1399	dd.10.48, wfu. 1.50, 'Red Eagle'.

Avro 683 Lancaster 7	G-ALVC	ex.	NX726	rd. 8.49, for spares use only.
Avro 685 York 1	G-AGNM	c/n	1215	dd.10.49, sold 11.52, trooping as XA192.
	G-AGNP	c/n	1218	dd. 9.52, sold 7.53, operated by Air Freight.
	G-AGNT	c/n	1221	dd. 9.52, sold 7.53, operated by Air Freight.
	G-AGNU	c/n	1222	dd. 9.52, sold 9.52, operated by Air Freight.
	G-AGNW	c/n	1224	dd. 3.54, sold 2.55.
	G-AGNY	c/n	1226	dd.10.49, sold 11.52.
	G-AGNZ	c/n	1227	dd.10.49, written off Berlin 24.8.52.
	G-AGOB	c/n	1229	dd. 5.54, sold 8.54.
	G-AMGK	c/n	1356	dd.12.50, sold 11.52, trooping as XA191.
	G-AMGL	c/n	1354	dd.12.50, sold 7.51, never entered service
	G-AMGM	c/n	1355	dd.12.50, sold 2.52.
Avro 652A Anson	G-AJBA	ex.	AX409	dd. 8.49, written off Paris 10.2.50.
Miles M65 Gemini 1A	G-AJKS	c/n	6289	dd. 7.56, sold 8.58.
Miles M14A Hawk Trainer Mk.3	G-AKPL	c/n	871	rd. 7.50, sold 3.53.
Auster J/1 Autocrat	G-AGXH	c/n	1898	rd. 1.51, sold 6.53.
	G-AIGM	c/n	2177	dd. 8.49, sold 3.53.
	G-AIJI	c/n	2307	rd. 1.51, sold 6.53.
Vickers 639 Viking	G-AGON	c/n	104	dd. 7.54, sold 1.55.
Vickers 657 Viking	G-AGRR	c/n	109	dd. 9.56, sold 4.57, 'Lord Charles Beresford'.
	G-AGRS	c/n	110	dd. 5.57, sold 8.59, 'Lord Charles Beresford'.
	G-AGRT	c/n	111	dd. 5.57, written off El Adem, Libya 26.2.58.
	G-AHOS	c/n	119	dd. 5.57, sold 8.57, 'Sir John Warren'.
Vickers 498 Viking	G-AHOR	c/n	118	dd.11.52, sold 3.53, never entered service.
	G-AHOT	c/n	121	dd.11.52, sold 3.53.
	G-AHOW	c/n	124	dd.11.52, sold 3.53.
Vickers 614 Viking	G-AHOZ	c/n	129	dd. 9.57, sold 5.58.
	G-AHPF	c/n	138	dd. 9.57, sold 4.58.
Vickers 643 Viking	G-AHPH	c/n	141	dd. 9.57, sold 3.58.
Vickers 610 Viking	G-AHPL	c/n	149	dd. 4.53, sold 6.53, never entered service.
	G-AHPM	c/n	152	dd. 4.53, written off Stavanger 9.8.61, 'Lord Rodney', trooping as XF632 and XG349.
	G-AHPO	c/n	157	dd. 4.53, wfu. 12.55, 'Lord Dundonald', later 'Sir Edward Howard'.
	G-AHPP	c/n	160	dd. 2.56, sold 2.56, 'Sir Charles Saunders'.
	G-AIVB	c/n	215	dd.12.55, sold 2.56, 'Robert Blake'.
	G-AIVC	c/n	216	dd. 6.54, sold 3.56, 'Lord Collingwood', trooping as WZ353.
	G-AIVH	c/n	221	dd. 9.54, wfu. 2.60, 'Lord Howe', trooping as XG896.
	G-AIVI	c/n	222	dd. 5.55, sold 3.56, never entered service.
	G-AIVL	c/n	225	dd. 4.55, wfu. 6.61, 'Lord Hawke'.
	G-AIVO	c/n	228	rd. 6.53, sold 6.62, 'Lord Vernon', trooping as XF630 and XG568.
	G-AJBN	c/n	240	dd.10.54, sold 12.55, 'Lord Nelson'.
	G-AJBO	c/n	241	dd. 6.53, written off Blackbushe 1.5.57, 'Sir John Benbow', trooping as XF629 and XF895.
	G-AJBP	c/n	242	dd.12.55, sold 12.58, 'Sir Edward Spragge', later 'Sir Edward Hughes'.
	G-AJBV	c/n	247	dd. 2.53, sold 5.53, never entered service.
	G-AJBW	c/n	248	dd. 6.55, sold 6.58, 'Sir William Cornwallis'.
	G-AJBX	c/n	249	dd. 6.53, sold 12.55, 'Sir Edward Hughes'.
	G-AJCD	c/n	255	dd.10.54, sold 6.62, 'Lord Barham', trooping as XF633 and XG350.
	G-AJCE	c/n	256	dd. 3.55, sold 4.56, 'Lord Exmouth'.
	G-AKBH	c/n	264	dd. .53, sold 4.61, 'Sir Henry Morgan', trooping as XG567.
Vickers 616 Viking	G-AIHA	c/n	146	dd. 7.56, wfu. 8.60, 'Sir Richard Kempenfelt'.
Vickers 618 Viking	G-AJPH	c/n	207	dd. 1.54, wfu. 10.61, 'Lord Dundonald', trooping as XJ804.
Vickers 635 Viking	G-AMGG	c/n	290	dd. 4.55, written off Agadir 12.12.59, 'Sir Robert Calder'.
	G-AMGI	c/n	297	dd. 6.55, wfu. 6.60, 'Sir Henry Harwood'.
	G-AMGJ	c/n	295	dd. 5.55, sold 4.57, 'Sir Charles Beresford'.
	G-AMNR	c/n	291	dd. 6.55, sold 2.56.
	G-AMNX	c/n	292	dd. 8.55, wfu. 10.61, 'Sir Philip Brooke'.
Vickers 621 Viking	G-APAT	c/n	153	dd. 5.57, sold 4.59, 'Lord Hood'.
Bristol 170 Freighter Mk.21	G-AHJD	c/n	12736	dd.10.56, sold 9.57.
	G-AIMA	c/n	12791	dd. 9.56, sold 6.57.
Airspeed AS65 Consul	G-AJXH	c/n	5167	dd. 3.56, sold 4.57.
Airspeed Oxford	G-AOUT	ex.	MP301	dd. .56, sold 8.56.
Douglas C47A Dakota	G-AJPF	c/n	13456	dd. 9.50, sold 1.51.
Douglas C47B Dakota	G-AGNG	c/n	26997	dd. 8.49, sold 1.51.
Douglas C47A Dakota	G-AGYZ	c/n	12278	rd. 3.54, sold 5.54.
	G-AHCT	c/n	12308	rd. 3.54, sold 5.54.
	G-AMPO	c/n	33186	dd. 5.54, sold 11.56, 'Sir Richard Kempenfelt'.
Douglas C47B Dakota	G-AMPS	c/n	32741	dd. 4.52, sold 5.53.
	G-AMPT	c/n	32935	dd. 5.52, sold 6.53.
	G-AMSO	c/n	33568	dd. 4.52, sold 11.52, never entered service.
Douglas C47A Dakota	G-AMVA	c/n	33163	dd. 9.52, sold 3.53.
Douglas C47B Dakota	G-AMYB	c/n	33346	dd. 1.53, sold 1.57, 'Lord Hood'.
Douglas C47A Dakota	G-AMZZ	c/n	33340	dd. 8.53, sold 3.54.
	G-AOJI	c/n	33156	dd. 5.56, sold 3.57, never entered service.
	G-AOYE	c/n	10028	dd. 1.57, sold 5.57, never entered service.
DH89A Dragon Rapide	G-AHKS	c/n	6812	dd. 7.55, sold 9.55.
	G-AJSL	c/n	6801	dd. 9.50, sold 1.52.
	G-AJXB	c/n	6530	dd. 6.55, sold 2.56.
	G-AKOB	c/n	6492	dd.10.52, sold 9.53.
Percival Proctor 1	G-AHES	c/n	H.6	dd. 1.51, sold 11.51.
Percival Proctor 5	G-AIAG	c/n	Ae.88	dd. 7.48, sold .54.
	G-AIER	c/n	Ae.99	dd. 8.49, sold 4.52.
Vickers 645 Valetta	G-APII	c/n	302	dd.12.57, wfu. 12.59, never entered service.
Vickers 651 Valetta	G-APIJ	c/n	491	dd.12.57, wfu. 11.59, never entered service.
DH114 Heron 1B	G-APKT	c/n	14019	dd. 2.58, sold 5.61.

DH114 Heron 1B	G-APKU	c/n 14025	dd. 2.58,	sold 10.61.	
	G-APKV	c/n 14045	dd. 2.58,	sold 9.58.	
	G-APKW	c/n 14046	dd. 2.58,	sold 8.59.	
DH104 Dove 5	G-AROI	c/n 04474	dd. 5.66,	ISOC ,	'Eaglet'.
	G-ASDD	c/n 04452	lsd.11.66,	ret. 12.66,	leased from Strathair.
Douglas DC6A	G-APOM	c/n 45519	dd. 8.58,	written off	Shannon 26.3.61.
	G-APON	c/n 45058	dd. 9.58,	sold 6.64.	
	G-APSA	c/n 45497	dd. 9.58,	sold 2.64.	
	G-ARMY	c/n 45457	dd. 3.61,	sold 4.64,	operated as VR-BBP 3.61.
Douglas DC6B	G-ARWJ	c/n 43844	lsd. 3.61,	ret. 12.62,	leased from CPA, operated as VR-BBQ 3.61 to 2.62.
	G-ARZO	c/n 45078	lsd. 6.62,	ret. 12.62,	leased from CPA.
Vickers 701 Viscount	G-ALWF	c/n 005	lsd.11.64,	ret. 5.65,	'City of Exeter', leased from Channel Airways.
	G-AMOA	c/n 009	lsd.11.64,	ret. 3.65,	'City of Newcastle', leased from Channel Airways.
	G-AMOC	c/n 013	lsd.12.63,	ret. 4.65,	'City of Glasgow', leased from Channel Airways.
	G-AMOE	c/n 017	lsd. 3.64,	ret. 12.64,	'City of Manchester', leased from Channel Airways.
	G-AMOH	c/n 021	lsd. 2.64,	ret. 3.65,	'City of Liverpool', leased from Channel Airways.
	G-AMOO	c/n 028	lsd. 5.64,	ret. 10.65,	'City of Birmingham', leased from Channel Airways.
Vickers 707 Viscount	G-APZB	c/n 030	lsd. 6.64,	ret. 11.64,	'City of Newcastle', leased from Aviation Overhauls.
	G-APZB	c/n 030	lsd. 1.65,	ret. 5.65,	'City of Manchester', leased from Channel Airways.
	G-ARKH	c/n 031	dd. 2.60,	sold 7.63,	operated initially as VR-BBJ.
	G-ARKI	c/n 032	dd. 1.60,	sold 7.63,	operated initially as VR-BBH.
Vickers 732 Viscount	G-ANRS	c/n 075	dd. 3.65,	ISOC ,	'City of Newcastle'.
Vickers 739 Viscount	G-ATDU	c/n 087	dd. 6.65,	ISOC ,	'City of Liverpool'.
Vickers 739A Viscount	G-ATDR	c/n 393	dd. 5.65,	ISOC ,	'City of Glasgow'.
	G-ATFN	c/n 394	dd. 8.65,	written off	Munich 9.8.68, 'City of Truro'.
Vickers 755 Viscount	G-AOCB	c/n 092	dd. .60,	sold 3.68,	'City of Edinburgh',operated as VR-BBL .60 to 9.62.
	G-AOCC	c/n 093	dd. .60,	sold 2.68,	'City of Belfast', operated as VR-BBM .60 to 9.62.
Vickers 805 Viscount	G-APDW	c/n 258	dd.12.57,	sold 6.59,	'Enterprise', operated as VR-BAX 4.58 to 6.59.
	G-APDX	c/n 312	dd. 3.58,	sold 2.60,	'Good Fortune', operated as VR-BAY 2.59 to 2.60.
Douglas C54E Skymaster	G-ASPM	c/n 10543	dd. 2.64,	sold 2.65.	
	G-ASPN	c/n 10337	dd. 2.64,	sold 2.65.	
	G-ASRS	c/n 27353	dd. 3.64,	sold 6.64,	never entered service.
BAC 1-11 207AJ	G-ATTP	c/n 039	lsd. 4.66,	ret. 12.67,	'Swift', leased from Zambia Airways.
	G-ATVH	c/n 040	lsd. 5.66,	ret. 12.67,	'Serene', leased from Zambia Airways.
BAC 1-11 301AG	G-ATPJ	c/n 033	dd. 6.66,	ISOC ,	'Stalwart'.
	G-ATPK	c/n 034	dd. 6.66,	ISOC ,	'Spur'.
	G-ATPL	c/n 035	dd. 7.66,	ISOC ,	'Superb'.
BAC 1-11 304AX	G-ATPH	c/n 110	dd. 4.67,	ISOC ,	'Salute'.
	G-ATPI	c/n 112	dd. 5.67,	ISOC ,	'Supreme'.
Boeing 707-465	G-ARWD	c/n 18372	dd. 2.62,	sold 9.62,	operated as VR-BBW 2.62 to 9.62.
	G-ARWE	c/n 18373	never	delivered ,	to have been operated as VR-BBZ.
Boeing 707-138B	G-AVZZ	c/n 17699	dd. 1.68,	ISOC ,	'Enterprise'.
	G-AWDG	c/n 17702	dd. 3.68,	ISOC ,	'Pheonix'.
Boeing 707-365C	G-ATZD	c/n 19590	dd. 2.68,	ISOC ,	'Enterprise', delivered as VR-BCP.
Bristol 175 Britannia 308F	G-ANCF	c/n 12922	dd. 1.64,	ISOC ,	'New Frontier'.
	G-ANCG	c/n 12923	dd. 1.64,	written off	Manston 20.4.67, 'Trojan'.
Bristol 175 Britannia 309	G-ANCH	c/n 12924	dd. 2.68,	sold 9.68,	never entered service.
Bristol 175 Britannia 312	G-AOVB	c/n 13230	dd.10.63,	ISOC ,	'Endeavour', converted to 312F.
	G-AOVC	c/n 13231	dd. 5.64,	ISOC ,	'Sovereign'.
	G-AOVE	c/n 13236	dd. 6.64,	sold 12.66,	'Perseverance', later 'Talisman', later 'Renown'.
	G-AOVF	c/n 13237	dd. 3.64,	ISOC ,	'Friendship', converted to 312F.
	G-AOVG	c/n 13238	dd. 4.64,	ISOC ,	'Bounteous'.
	G-AOVH	c/n 12925	dd.12.64,	sold 3.65,	'Crusader'.
	G-AOVK	c/n 13419	dd. 8.65,	ISOC ,	'Concord'.
	G-AOVL	c/n 13420	dd. 7.65,	ISOC ,	'Resolution'.
	G-AOVM	c/n 13421	dd. 3.64,	ISOC ,	'Team Spirit', converted to 312F.
	G-AOVN	c/n 13422	dd. 6.64,	ISOC ,	'Prospect'.
	G-AOVO	c/n 13423	dd. 1.64,	written off	Innsbruck 29.2.64, 'Bonaventure'.
	G-AOVR	c/n 13429	dd. 3.65,	sold 10.66,	'Talisman'.
	G-AOVS	c/n 13430	lsd. 8.65,	ret. 12.65,	leased from Lloyd International.
	G-AOVT	c/n 13427	dd. 9.63,	ISOC ,	'Enterprise', later 'Ajax'.
Bristol 175 Britannia 313	G-ARXA	c/n 13234	dd. 5.66,	ISOC ,	'Talisman', later 'Renown'.
Bristol 175 Britannia 314	G-ASTF	c/n 13453	lsd. 5.64,	ret. 10.64,	'Concord', leased from CPA.
Bristol 175 Britannia 318	G-APYY	c/n 13432	lsd. 3.60,	ret. 10.61,	leased from Cubana.
Bristol 175 Britannia 319	G-AOVA	c/n 13207	dd. 4.64,	ISOC ,	'Justice'.
Bristol 175 Britannia 324	G-ARKA	c/n 13516	dd. 3.61,	ISOC ,	'Good Fortune'.
	G-ARKB	c/n 13517	dd. 5.61,	ISOC ,	'Endeavour', later 'Resolution'.

BRITISH ISLAND

Airlines (Jersey) Ltd.
British Island Airways Ltd.
British United (Channel Islands) Airways Ltd.
British United Island Airways Ltd.
Morton Air Services Ltd.

Bases: Blackpool and Jersey Airports.

British United (Channel Islands) Airways was formed on November 1 1962 by merging the passenger and cargo route networks of Silver City Airways and Jersey Airlines, both now members of the Air Holdings group. The airline operated a large fleet of Dakotas, Heralds, Herons, Doves and Viscounts mainly on scheduled services to the Channel Islands and the Isle of Man. In addition, the airline also flew nightly newspaper services to Germany and the Channel Islands formerly flown by Transair. The airline's scheduled services were based on Jersey and Blackpool, but although the airline took over the Gatwick to Guernsey route from its parent company, the Gatwick to Jersey service continued to be operated by British United Airways. Jersey Airlines had become part of the British United Airways group on May 20 1962, and because its name was such a good advertisement for its services, the name of Jersey Airlines was retained, and used throughout the 1962 and 1963 seasons, but on August 1 1963 it changed its name to British United (Channel Islands) Airways.

The Manx division of the airline served Blackpool, Leeds, Dublin, Belfast, Glasgow and Newcastle from the Isle of Man, while the Channel Islands division served several points from Jersey, Guernsey and Alderney, and on June 2 1963 Herald G-APWI inaugurated a Newcastle to Jersey service, which was later flown by Viscounts, but was dropped from the network at the end of 1966. During 1963, the airline's Heralds flew the British United Gatwick to Rotterdam service, and on April 1 1963 the airline opened a Gatwick to Blackpool service. This was flown with 32 seat DC3s, and operated thrice weekly. Initially load factors were in the thirty per cent bracket, but these quickly doubled by the time that an additional two flights a week were introduced in May. The flights were timed to connect with continental flights from Gatwick, but on September 30 the service was dropped and taken over by Autair.

During the next five years, there were few changes, although the Dakota fleet was dwindling as the aircraft were being sold, and two more Heralds were acquired at the end of 1966. Some of the Dakotas, Herons and Doves were transferred to Morton Air Services in 1965 who were operating scheduled services to Alderney, Swansea and Exeter from Gatwick. Maintenance was concentrated at Jersey and Blackpool, and rail-air services were operated from London to the Channel Islands, using Southampton as the mainland airport. Southampton proved ideal in this respect, with a railway station at the airport, and passengers could quickly make the change from a train to the aircraft. After lengthy disscussions and threats of industrial action, a new airline named British United Island Airways was formed on November 1 1968. Prior to this, it had been proposed that a new airline named Channel Islands and Manx Air Services (CIMAS) be formed but this proposal was not adopted. British United Island Airways was composed of British United (C.I.) Airways, British United Airways (Manx) and Morton Air Services, and all three airlines now adopted this new title. Services, though, continued normally although the Morton services from Swansea were terminated, and in December 1969 the airline announced that it was to suspend its services to Alderney from Southampton and Gatwick. Also dropped were their services between Jersey and Dinard, St.Brieuc and Quimper, and when services ceased on December 31 1969, they were immediately taken over by Aurigny Air Services. British United Island Airways now operated no services to Alderney, and therefore the requirement to operate Heron and Dove aircraft no longer existed, and these aircraft were sold.

On July 20 1970, with the prospect of British United Airways being sold, British United Island Airways changed its name to British Island Airways. At the same time a new colour scheme was adopted for its aircraft, and when later in the year British United joined with Caledonian Airways, British Island Airways was not part of the deal, and remained in the ownership of the British and Commonwealth Shipping Group. On April 1 1971, British Island Airways opened a Gatwick to Antwerp service, flown in conjunction with SABENA, using Heralds, and in addition the airline flew Antwerp to Heathrow services for SABENA. In May 1973, the airline restarted services from Hurn to the Channel Islands which had not been flown for several years by the airline. Heralds were used on passenger flights to Jersey and Guernsey, and Dakotas were used for all-cargo flights between Bournemouth and the Channel Islands. Although these flights were suspended during the following winter, they were resumed again in the spring of 1974. On November 1 1973, British Island Airways opened a twice daily service between Gatwick and Manchester flown on behalf of British Caledonian Airways. This route is flown by Heralds, and is timed to connect with British Caledonian's scheduled services radiating from Gatwick. On May 30 1974 British Island operated its last Dakota flight. For several years the Dakotas had been relegated to all-cargo work, and on May 30 G-AMSV touched down at Bournemouth after completing its last scheduled cargo flight from Guernsey. The aircraft was carrying a cargo of flowers, and later positioned to Gatwick, and then onto Blackpool to await sale to a new owner.

British Island Airways was thus now an all-turbine operator flying a dozen Heralds on scheduled services throughout the United Kingdom. Little charter work is undertaken, but the Heralds do operate the B.C.A.L. services from Gatwick to Manchester, and also the Silver Arrow service from Gatwick to Le Touquet. During 1975 British Island Heralds did undertake some charter work and airports visited while engaged on this form of work included Bristol, Humberside, Liverpool and Prestwick. Nightly newspaper services from Gatwick are now run to Jersey, Guernsey, Belfast, Hanover, and Dusseldorf, and the airline has also recently taken over the former Cambrian night freight run from Liverpool to the Isle of Man. In October 1975 British Island Airways obtained a contract from the Irish Press to transport newspapers from Dublin to Luton on Saturday nights - the service had previously been flown by Alidair's Viscounts. British Island's normal scheduled services link the Channel Islands with Southampton, Bournemouth, Exeter, Blackpool, Gatwick (Guernsey only) and Paris; and Exeter with Southampton, Dublin and Belfast; and the Isle of Man with Blackpool, Edinburgh, Leeds, Glasgow, Dublin and Newcastle. British Island Airways now carries over half a million passengers on its services annually, and the airline still remains as an almost wholly owned subsidiary of the British and Commonwealth Shipping Group. This group has a ninety one per cent interest in British Island, while the remainder of the shares are held by the Eagle Star Insurance Group. For the future, British Island Airways is finding it increasingly difficult to obtain further Heralds, and therefore the acquisition of a new type of aircraft is extremely probable in the near future.

Fleet:

Douglas C47 Dakota	G-AKNB	c/n 9043	ISOF	sold 2.69.
Douglas C47A Dakota	G-ALPN	c/n 12218	ISOF	sold 11.68.
	G-AMHJ	c/n 13468	ISOF	sold 1.74.
Douglas C47B Dakota	G-AMJU	c/n 25934	ISOF	sold 3.68.
	G-AMPZ	c/n 32872	ISOF	sold 11.69.
	G-AMRA	c/n 26735	ISOF	sold 1.74.
	G-AMSV	c/n 32820	ISOF	sold 9.74.
	G-AMWV	c/n 25600	ISOF	sold 1.68.
	G-AMYJ	c/n 32716	ISOF	sold 5.69.
	G-AMYV	c/n 32943	ISOF	sold 10.67.
	G-AMYX	c/n 33042	ISOF	sold 8.66.
	G-AMZF	c/n 27078	ISOF	sold 6.66.
	G-AMZG	c/n 33416	ISOF	sold 6.66.
	G-AMZH	c/n 27110	ISOF	sold 5.65.
	G-ANAE	c/n 26101	ISOF	sold 4.68.
	G-ANEG	c/n 33444	ISOF	sold 4.65.
	G-ANTB	c/n 27207	ISOF	written off Jersey 14.4.65.
	G-ANTC	c/n 26111	ISOF	sold 10.68.
Douglas C53D Dakota	G-AOBN	c/n 11711	ISOF	sold 11.71, calibration aircraft.
Douglas C47B Dakota	G-AOUD	c/n 25573	ISOF	sold 1.69.
DH104 Dove 1B	G-AMYO	c/n 04086	ISOF	sold 7.69.
	G-ANAN	c/n 04062	ISOF	sold 6.69.
DH104 Dove 6	G-AOYC	c/n 04065	ISOF	sold 9.69.
DH114 Heron 1B	G-ANSZ	c/n 14047	ISOF	sold 7.70.
	G-ANWZ	c/n 14081	ISOF	sold 7.70.
	G-AOXL	c/n 14015	ISOF	sold 7.70.

DH114 Heron 2	G-AOGO	c/n	14096	ISOF , sold 1.69.
Vickers 708 Viscount	G-ARBY	c/n	010	ISOF , sold 2.67.
	G-ARER	c/n	012	ISOF , sold 6.66.
	G-ARGR	c/n	014	ISOF , sold 9.66.
Vickers 736 Viscount	G-AODG	c/n	077	ISOF , sold 1.67.
Vickers 831 Viscount	G-APND	c/n	402	ISOF , sold 1.69.
	G-APNE	c/n	403	ISOF , sold 4.67.
	G-ASED	c/n	419	ISOF , sold 2.67.
Handley Page Herald 201	G-APWE	c/n	153	ISOF
	G-APWF	c/n	154	ISOF
	G-APWG	c/n	155	ISOF
	G-APWH	c/n	156	ISOF
	G-APWI	c/n	157	ISOF , sold 11.68.
	G-APWJ	c/n	158	ISOF
Handley Page Herald 209	G-BAZJ	c/n	183	dd. 6.73
Handley Page Herald 203	G-ASBG	c/n	164	dd. 7.73
	G-BBXI	c/n	184	dd.12.73
	G-BBXJ	c/n	196	dd.12.73, written off Jersey 24.12.74.
Handley Page Herald 207	G-ATHE	c/n	165	lsd. .65, ret. .66, leased from Handley Page Ltd.
Handley Page Herald 210	G-AVEZ	c/n	169	dd.12.66
Handley Page Herald 211	G-ASKK	c/n	161	dd.12.66
Handley Page Herald 213	G-AYMG	c/n	179	dd.11.70
	G-AVPN	c/n	176	dd. 7.73
Handley Page Herald 214	G-ATIG	c/n	177	lsd. 3.75, ret. 10.75, leased from British Midland.

BRITISH MIDLAND

British Midland Airways Ltd.
Derby Airways Ltd.
Derby Aviation Ltd.
Wolverhampton Aviation Ltd.
Air Schools Ltd.

Bases: Initially Burnaston (Derby) and Wolverhampton, later the East Midlands Airport (Castle Donington).

British Midland Airways' history can be traced back to 1938 when, in October of that year, Air Schools opened a reserve flying school at a new airport at Burnaston, Derby. During the war, flying training continued, training pilots and navigators for the Royal Air Force, and a further flying school was soon established at Wolverhampton. The original reserve flying school was formed by Wing Commander N.Roy Harben and was initially equipped with a fleet of Hawker Hinds as No.16 Reserve Flying School. The Derby Aero Club was also set up by Air Schools before the outbreak of war in 1939, and during the war the reserve flying school operated a fleet of Miles Magisters. Shortly after the outbreak of war, another reserve flying school was formed by Air Schools at Wolverhampton Airport, and from here the company operated a fleet of Tiger Moths on preliminary flying training. This reserve flying school was No.25 Reserve Flying School, and when the war came to an end in 1945 both Reserve Flying Schools remained in existence.

By 1945 both schools were flying a large number of Tiger Moths on ab initio training, and shortly after the end of the war the Derby based organisation replaced these aircraft with Chipmunks, Prentices and Ansons, while the Wolverhampton school re-equipped with Chipmunks. The Ansons were used as navigational trainers, and at Wolverhampton the company also looked after five Tiger Moths of the Durham and Leeds University Air Squadron. In addition to the aircraft flown by these reserve flying schools, Air Schools also operated a large fleet of light aircraft on behalf of the Derby and Wolverhampton Aero Clubs, and these aircraft were often used for charter flights. The company started charter operations on August 21 1947, and aircraft used included Austers, Messengers and Geminis. By 1949 Air Schools was responsible for the operation of nine Tiger Moths and two Ansons at Derby, and for twelve Tiger Moths and two Ansons at Wolverhampton. The replacement of these aircraft by the more modern training types took place over the next two years.

Early in 1949 the company decided that it would like to expand its commercial charter operations, and on February 16 1949 Derby Aviation was registered as an airline company by Group Captain Clifford A.B.Wilcock and Edward W.Phillips. Still based at Burnaston, this company started ad-hoc passenger and freight charter operations during the spring with a Rapide and a Miles Aerovan aircraft which had been bought from Kenning Aviation during 1948. On June 29 1949 the Wolverhampton Flying School changed its name to Wolverhampton Aviation, and thus two new airlines had been born. In addition to the fleets owned by the two companies, both Derby Aviation and Wolverhampton Aviation used aircraft on loan from their parent company, including Geminis, Messengers, Austers and a Miles Monarch. Early work for Derby Aviation's Rapide was the operation of a large number of Army Co-operation flights under contract to the Army. However, this aircraft did find time to operate other charter flights, and both Derby Aviation's and Wolverhampton Aviation's aircraft became very familiar sights at the many air displays in the Midlands during the early post-war years. At these displays, the companies' aircraft were frequently used for joy riding, particularly at the air shows held at Wolverhampton and Derby.

One of Derby Aviation's founders, Group Captain Wilcock, was an influential Midlands' personality. At this time, he was a member of Parliament for Derby, and in addition to being Chairman of Air Schools, Derby Aviation and Wolverhampton Aviation, he was also a director of Skyways, and during the Second World War he acted as Senior Personnel Officer of Royal Air Force Transport Command. For the summer of 1949, Derby Aviation operated a Dragon Rapide and a Miles Aerovan on ad-hoc passenger and freight charter flights, pleasure flying at Derby and Wolverhampton, flying training and army co-operation flights in the Midlands. The many ad-hoc passenger and freight flights took the Rapide to the Channel Islands, the Isle of Man and to many race meetings around the country. The Aerovan was replaced by a Rapide early in 1950 and this aircraft was based at Wolverhampton and flown by Wolverhampton Aviation. Meanwhile, Derby Aviation was also making a name for itself as a fully A.R.B. approved maintenance and repair organisation - apart from the conversion of many military aircraft to civilian standards, Derby Aviation also converted well over a dozen Miles Geminis to Gipsy Major power. Flying training was still popular at Derby and Wolverhampton, and with the acquisition of the Elstree Flying Club in the early summer of 1952, Air Schools increased its civilian flying club fleet to twenty one aircraft. However, early in 1953 both reserve flying schools run by Air Schools at Derby and Wolverhampton were closed down. Derby Aviation still continued to maintain Harvards and Chipmunks operated by several Midland based University Air Squadrons, but all flying on behalf of the Royal Air Force came to an end. The closure of these schools led Derby Aviation to turn to the operation of more charter services - in addition to the Rapides at Derby and Wolverhampton, the company also offered the Austers of the Elstree Flying Club for charter. During 1952, with a staff of fifty five people including four pilots, Derby Aviation flew forty six thousand revenue miles with its Rapide on charter flights, and this prompted Derby Aviation to apply for a scheduled passenger service linking the Midlands with the Channel Islands.

On July 18 1953 one of the Rapides opened a scheduled passenger service from Derby, via Wolverhampton to Jersey. This summer only service was suspended at the end of September after over twelve hundred passengers had used the service - 573 passengers from Wolverhampton, and 634 from Derby. The average load factor for the season was a credible seventy five per cent, and with the continued operation of army co-operation flights, more Rapides were acquired for the 1954 season. For the 1954 summer season the Rapides flew schedules to the Channel Islands from Birmingham, Derby, Nottingham and Wolverhampton. These services were all flown on Saturdays and Sundays from April to September inclusive and during the summer of 1954 these services proved to be very popular

British United Airways' Dakota G-AMJU at Newcastle on June 2 1963.

Ian MacFarlane

Derby Airways' Dakota G-AOGZ at Newcastle on June 5 1962

Ian MacFarlane

British Nederland's second Dakota G-AJZX operated seventy six sorties on the Berlin Airlift.

A J Jackson

Douglas DC-6A G-APNO in British United's interim colour scheme at Heathrow Airport on August 20 1960.

R A Scholefield

Another of British United's interim colour schemes is displayed here on Viscount G-AOXV at Newcastle on September 7 1960.

Ian MacFarlane

British United Airways' Bristol Freighter 31 G-ANMF at Newcastle on July 2 1967.

Ian MacFarlane

This fine photograph of British United's Britannia G-APNA was taken at Southend Airport on October 11 1966.

Transportraits/John Goring

British Westpoint's Rapide G-AHKU was one of three former BEA aircraft acquired in May 1964, and pictured here at St. Mawgan in October 1965.

Transportraits

British Westpoint's Dakota G-ALYF was the aircraft used for the Air France services to Lille - look closely, and the stickers are visible behind the door.

Ian MacFarlane

with numerous holidaymakers. However, 1954 also saw the start of a rundown of the maintenance organisation at Wolverhampton and gradually all overhaul and repair work was transferred to Derby. In addition, the name of Wolverhampton Aviation disappeared at the end of the 1954 season and all operations were henceforth conducted under the name of Derby Aviation.

In April 1955 Derby Aviation acquired a Douglas Dakota for use on its popular summer services to the Channel Islands, and on May 1 1955 G-ANTD went into service on the route from Burnaston and Birmingham to Jersey. The services from Wolverhampton and Nottingham to the Channel Islands continued to be flown by Rapides and during 1955 a service to Ostend was also operated from Burnaston. Two Miles Marathons were bought late in 1955 for use on Derby Aviation's scheduled services, and during the spring of 1956 these aircraft entered service on the routes to Jersey and Ostend, and also on two new services from Derby to Glasgow and the Isle of Man. The Marathons were operated as eighteen or twenty seaters and as they were not equipped with de-icing equipment they were only operated during the summer months. At the end of the 1956 season the services from Wolverhampton and Nottingham were dropped completely, and Derby Aviation's main base was now Burnaston Airport.

While Derby Aviation had been busy developing its passengers services, the other sides of the business had also been actively engaged in various activities. An Anson (G-AMDA) had been acquired in July 1955 and after conversion this aircraft, carrying an airborne magnetometer for geological survey work, was operated on survey work in the Midlands.The aircraft was flown in conjunction with Canadian Aero Services(whose titles the aircraft carried) and was sponsored by the Nuffield Foundation. This aircraft, flown by Captain R.Wallis, flew around the Midland countryside at a height of 1,800 feet recording variations in the earth's magnetic field. Other survey aircraft converted by Derby Aviation included a batch of ten Mosquito PR.35s which were converted at Bunaston for Spartan Air Services in Canada. These aircraft were ferried from Silloth and Shawbury to Burnaston by the manager of the Elstree Flying Club, Mr.David Ogilvy, and after conversion were then flown to Prestwick. At Prestwick the Mosquitos were then handed over to Mr.R.W.Nook of West London Air Charter who then completed the ferry flight to Canada.

Derby Aviation's operations were, therefore, extremely varied and in addition to those services already mentioned, the airline undertook a number of charter flights, particularly during the winter months when work was slack. By the end of 1956 Derby Aviation had two Dakotas in service, and when the Hungarian revolution took place at the end of the year, these two aircraft were immediately pressed into service on the airlift of refugees between Linz, Vienna and Blackbushe. The two Dakotas carried blankets and medical supplies on their outbound trips to Austria, and on their return flights refugees were carried accompanied by a Red Cross escort. By the time the airlift ended on December 14 1956, Derby Aviation had operated eighteen round trips bringing well over six hundred people out of Austria. Other charters undertaken by Derby Aviation included many on behalf of Rolls Royce which took the Dakotas as far away as India carrying engines, spare parts and technicians. By 1957,the survey division had been renamed Derby Aero Survey, and its fleet, consisting of three Austers and an Anson, had moved to Elstree. A second Anson was also later acquired, and later these aircraft were used by Air Schools at Elstree for flying training and navigational instruction. Eventually the company became known as the London School of Flying before it was sold to Polyfoto Air Taxis.

Three Dakotas were in service for the 1958 summer season and these aircraft operated inclusive tour charter flights on behalf of several Midlands based tour operators. These services were mainly flown by the Dakotas and linked Derby with Austria, Switzerland, France, Italy, Majorca and the Costa Brava. During the summer, scheduled services were flown from Derby to Glasgow, the Isle of Man, Jersey and Ostend, while a service to Antwerp was flown via Birmingham. In addition, the Marathons opened scheduled services to Jersey from Northampton and Oxford - the Marathons proved to be ideal for these services with their good airfield performance from these grass airfields.

On March 12 1959 Derby Airways was registered as an airline company, and took over all the services and aircraft previously flown by Derby Aviation. In May 1959 the airline opened a Marathon service from Staverton to Guernsey and Jersey which was flown during the summer months. For the 1959 season, Derby Airway's scheduled service network included Birmingham, Cambridge, Derby, Glasgow, Guernsey, Jersey, Luton, Oxford, Ostend and Staverton. The inclusive tour side of the business was rapidly building up with departures from Birmingham, Burnaston, Gatwick and Manchester to France, Italy, and Spain during the summer months, and to Austria, Norway and Switzerland during the winter months. From Manchester, Derby Airways operated a once weekly inclusive tour charter flight, on Fridays, from Manchester to Ostend by Dakota. For the summer of 1960 a fleet of six Dakotas was in service with Derby Airways, and on May 31 1960,one of these aircraft opened a Cardiff to Palma, service via Bristol. Later, services were also opened from Cardiff to Luxembourg, Nice, Perpignan and Ostend again using Dakotas. During 1960 the Marathons were regular visitors to Hurn Airport on their many runs to the Channel Islands,and when the last Marathon was withdrawn from service at the end of September 1960, Derby Airways used Overseas Air Transport Herons for many of its services to the Channel Islands. During the year the airline applied to the A.T.L.B. for permission to use leased Overseas Air Transport Herons on a wide variety of services including services to the Channel Islands from Derby, Birmingham, Nottingham, Northampton and Wolverhampton. In addition, services from Derby, Luton and Northampton to the Isle of Man were applied for, and also services to Ostend from Southend, Derby, Luton, Nottingham and Birmingham, and to Jersey from Bournemouth, Oxford, Staverton, Luton and Cambridge. Although many of these services were never flown by Overseas Air Transport's Herons, these aircraft certainly did appear on services from Cambridge, Bournemouth and Derby to the Channel Islands late in 1960 and during 1961.

By 1960 Derby Airways was carrying over twenty thousand passengers on its scheduled domestic and international services with many more people being carried on its inclusive tour charter flights. Two more Dakotas were added to the fleet for the 1961 season bringing Derby's Dakota fleet to eight aircraft, but one of these aircraft was lost on October 7 1961 in Derby Airway's first fatal accident. The Dakota had left Gatwick Airport early in the evening on a charter flight to Perpignan with thirty one passengers and three crew aboard. The passengers were a party of local government officials, civil servants and their families who were tarvelling to Perpignan to spend holidays in the South of France organised by Whitehall Travel. However, the aircraft crashed into the side of Mount Canigou at a height of 7,500 feet and all on board were killed. In October 1961, Derby Airways acquired three Argonauts from the liquidator of Overseas Aviation, and in May 1962 the first of these aircraft entered service on scheduled services from Cardiff and Bristol to Palma, and on inclusive tour services from the Midlands to Austria, France, Italy and Spain. During the winter of 1962 these aircraft were also used for inclusive tour charter flights to Tenerife in the Canary Islands.

On April 2 1963, Dakota G-AMSX opened a Derby - Birmingham - Newcastle scheduled service which was subsequently flown thrice weekly in each direction. This route was acquired from Executive Air Transport from whom Derby also acquired a Birmingham to Sandown service which was also later opened by Derby. Close association had been maintained with Executive Air Transport and their Dakota (G-ANEG) had often been used by Derby Airways on their services from Burnaston to Glasgow. During 1963, Derby Airways also operated services from Carlisle to Gatwick and Belfast,and Mercury Airline's Heron service from Manchester to Newcastle was also flown in association with Derby Airways. During the 1963 summer season the Argonauts operated many inclusive tour flights including services from Manchester to Genoa, Ostend, Klagenfurt and Perpignan, while from Bristol and Cardiff services were flown to Alicante, Barcelona, Palma and Tarbes, with a Dakota service from Swansea, Cardiff and Bristol to Ostend. From Birmingham, the Argonauts flew inclusive tours to Barcelona, Cork, Ostend, Palma, Perpignan, Tarbes and Venice, but Dakotas still operated some of the services to Ostend.

During the winter of 1963, Derby found that with the severe fall off in passenger traffic, it was forced to find other work for its aircraft, and with its convertible passenger/cargo Dakotas, Derby Airways opened a Derby to Belfast freight service. Further freight services to Amsterdam from Birmingham, Bristol, Cardiff and Derby were also projected, but not for introduction until May 1 1964. Inclusive tour charter operations from Birmingham, Bristol, Cardiff, Derby and Manchester continued during 1964 and on July 30 1964 it was announced that Derby Airways was to change its name to British Midland Airways. This name change officially took place on October 1 1964, and this was followed on February 1 1965 by the introduction of the airline's first turboprop aircraft - a Handley Page Dart Herald. Prior to this, in October 1964 British Midland Airways took over the former Mercury Airlines routes from Liverpool and Leeds to Exeter and Sandown, and at the end of the month the airline also took over the service from Manchester to Newcastle. However, this latter service was soon suspended, and although restarted on April 1 1966, the service was again closed down on June 30 1966 due to poor support.

In April 1965, British Midland moved to its present base at East Midlands Airport at Castle Donington which had just opened - British Midland becoming Castle Donington's first airline residents. During April new services from Castle Donington to Palma and Gerona were inaugurated, and in 1966 a service to Newquay was also opened. In January 1967, the first Viscount was introduced, and by June several aircraft of this type were in service operating on virtually all British Midland's routes apart from the services from Manchester and East Midlands to Ostend which were still flown by Argonauts, and the flights to Sandown from Birmingham, East Midlands and Manchester which were operated by Dakotas. However, 1967 was not a happy year for British Midland - on June 4 one of the Argonauts (G-ALHG) crashed while on approach to Manchester Airport while inbound from Palma with a full load of inclusive tour passengers. Power was lost from two engines some distance from touch-down, and being unable to maintain height, the aircraft crashed in Stockport, tragically killing almost all those on board. The other Argonauts were grounded, and one aircraft departed for Boscombe Down where extensive tests were carried out on the aircraft to determine the cause of the accident. Although this aircraft later returned to Castle Donington, the Argonaut's career with British Midland had come to a sad and abrupt ending and the two surviving aircraft were later scrapped.

In February 1969, Minster Assets, who now controlled both British Midland and Invicta Airways, brought the two airlines together into a merger. From this merger, it was hoped to form a new airline named Air Britain, but this merger only lasted a matter of weeks before Hugh Kennard repurchased the cargo part of the airline, which had been renamed British Midland - Invicta Air Cargo, and subsequently British Midland continued operations normally, except that it had acquired Invicta's two Viscounts in the reshuffle.

When Autair International announced during 1969 that it would be suspending all its scheduled services on October 31, British Midland immediately applied for several of Autair's services including those from Teesside. The airline was awarded several of Autair's services from Teesside, and on November 1 1969 British Midland opened its Teesside to Heathrow service. This service was flown initially by Viscounts, but on May 4 1970 the Viscount services were replaced by a thrice daily jet service between the two airports flown by the airline's newly acquired BAC 1-11s. 1970 proved to be a year of change for British Midland - a Boeing 707 was acquired in April 1970 for use on transatlantic charter flights from Stansted, Gatwick, Prestwick and East Midlands to North America and Canada. The Boeing was handed over to British Midland at Windsor Field, Nassau, on April 15 and arrived at Castle Donington on April 26. On the following day the aircraft left Castle Donington for crew training at Stansted, and after the award of British Midland's U.S. Foreign carrier's permit early in May, the 707 operated its first revenue service on May 3 with a charter flight from Castle Donington via Keflavik to North America. In addition to these affinity group charters to North America, the aircraft was also used for charters to the Caribbean, the Far East and Africa, and also on some of British Midland's tour and scheduled services, including the Manchester to Ostend service when loads were heavy. In addition to introducing 707s during 1970, British Midland also accepted delivery of three BAC 1-11s for use on both the inclusive tour and scheduled service networks. On February 23 1970, the first 1-11 (G-AXLL) operated a proving flight from Castle Donington to Jersey, and on the following day the aircraft entered service on this route. Over the next six weeks the 1-11s were gradually introduced onto British Midland's services from Castle Donington to Dublin, Glasgow and Luton. The Teesside to Jersey service was inaugurated on May 23 1970 using a Viscount, and on the same day a Teesside to the Isle of Man service was also opened. The Isle of Man also had British Midland services to Birmingham, Coventry and East Midlands, and services to Newquay were flown from Birmingham, East Midlands, Glasgow and Manchester during the summer months.

With the 1-11s, British Midland was able to expand further into the inclusive tour charter market, and apart from the local tour operators, the airline also flew services for Vista Jet, Clarksons, Horizon Holidays, Global and many others. From Luton the 1-11s flew to Gerona, Ibiza, Alicante, Palma and Malaga; from Glasgow to Alicante, Gerona, Ibiza and Palma; from Manchester to Alicante, Gerona, Ibiza and Palma; and from Belfast to Gerona and Palma. In addition, a once weekly service was operated from Bristol to Palma using Viscounts, although later British Midland's 1-11s took over the Courtline services from Lulsgate, when that airline with-drew from operating services to Bristol in 1972. However, British Midland did not find its jet operations to be totally successful - with the 1-11s, the airline found that by 1972 the inclusive tour charter market was becoming so cut throat that it was more economical to withdraw from this market altogether, rather than try to compete. Subsequently, one aircraft was leased to Courtline, one aircraft was sold, and the third aircraft was retained for scheduled services, and for a limited tour programme from Bristol, Castle Donington and Luton. However, this aircraft was also sold, in April 1974, and British Midland stopped all inclusive tour charter flying. The transatlantic charter market had also become tough, and even with all 189 seats filled on many services the 707s were still making little money, but rather than sell these aircraft, British Midland looked for other work for its two aircraft. British Midland's prayer's were answered when, on November 11 1972, the airline's first 707 (G-AYBJ) started flying Sudan Airways' Blue Nile scheduled service from Heathrow to Khartoum under contract to Sudan. The second 707 entered service in Sudan Airways' colours on December 15, and this two year contract was worth £3.3 million to British Midland. This contract almost fully occupied the two 707s, although at weekends during the summer the aircraft did occasionally appear at Prestwick operating ABC charters to the United States and Canada.

When Channel Airways suspended operations on February 29 1972, British Midland immediately applied to the A.T.L.B. for permission to take over their services to the Channel Islands from Hurn, Southend and Stansted. Pending a decision by the board, British Midland was allowed to take over the services temporarily and on March 3 1972 the first services between Southend, Bournemouth, Jersey and Guernsey were flown by Viscount G-APNE. On April 10, Viscount G-AZLT inaugurated the East Midlands and Birmingham to Brussels and Frankfurt service, and during June it was announced that British Midland had been granted a Stansted and Southend to Jersey and Guernsey licence effective from July 1 1972, but that the Bournemouth to Channel Islands licence had been awarded to Dan-Air Services, and Dan-Air took over this service using 748s. On April 1 1973, a daily East Midlands to Paris service was opened using Viscounts, followed on April 30 by the opening of a Newquay to Heathrow service, again using Viscounts. Later the same day, the same Viscount inaugurated the airline's twice weekly London to Strasbourg scheduled service flown in competition with Air France.

In March 1973 the first of three former SADIA Heralds arrived at Castle Donington for British Midland Air-ways. This aircraft (G-ASVO) left Sao Paulo on March 14 and after flying via Brasilia, Carolina, Belem, Cayenne, Barbados, San Juan, Nassau, Miami, Richmond, Bangor, Monkton, Goose Bay, Narsarsuaq, Keflavik and Abbotsinch, the Herald arrived at Castle Donington on March 23 after a total flying time of almost forty eight hours. The introduction of the three Heralds into service was not particularly swift - the three aircraft spent many weeks on the apron at Castle Donington resting after their long journey from South America, but eventually G-BAVX was wheeled into British Midland's hanger for conversion. This aircraft flew again in October, and on November 12 1973 it operated its first service for British Midland when it operated a Newquay to Heathrow scheduled service. During the winter of 1973, the Heralds operated the service to Newquay, and they also appeared on other British Midland scheduled services from Castle Donington to Dublin, Glasgow and Belfast. Two of the Heralds were later based at Aberdeen for much of 1974, in order to undertake charter flights on behalf of the many oil companies in the area. British Midland's first oil rig support flights were flown in January 1974 when two Viscounts were flown under contract to Site Aviation from January 14 onwards. However, this contract came to an early end in April 1974, when Site Aviation ceased operations. Since then, though, British Midland has based one of its Viscounts at Aberdeen and this aircraft has flown charters to other Scottish airports and also to Scandinavian airports.

The contract with Sudan Airways came to an end in July 1974, and on July 2 G-AYVE returned to Castle

Donington where it was resprayed in British Midland's colours, and later entered service with British Caledonian on July 10. The other 707 (G-AYBJ) returned to Castle Donington on July 30, and then departed to Heathrow on lease to Iraqi Airways on August 10 1974. The return of these two aircraft to British Midland and the airline's surplus Heralds and Viscounts prompted the airline to form a leasing division to arrange the leasing out of these aircraft to other operators with all crews and ground support arranged or supplied by British Midland. The success of this operation was soon made very evident, and within a year of the opening of this leasing division the 707s have flown services for Kuwait Airways, British Airways, British Caledonian Airways, Zambia Airways, East African Airways, Sterling Airways, Iraqi Airways, El Al, and Nigeria Airways. Then on May 28 1975 G-AYVE flew from Stansted to Damascus to start flying scheduled services on behalf of Syrian Arab Airlines from June 1 onwards. Three aircraft are now engaged on this work, and with other contracts flowing in three more Boeing 707s were obtained during 1975 to cope with the massive volume of work. During the summer of 1975, in addition to the airlines already mentioned, the 707s flew services for Sudan Airways, Dan-Air, Bangladesh Biman, Condor, Royal Air Maroc and Air Inter and on September 13 G-AYVG left Stansted on a six month lease to Pakistan International, while on September 27 G-AZWA started a six month contract with Malaysian Airlines. The Boeing 707s have been known to operate services for British Midland - on April 5 1975 G-AYVE operated a charter flight from Heathrow to Saigon to collect eighty Vietnamese orphans who were then flown back to Heathrow on April 6. This flight was much publicised at the time, and was probably one of the most historic flights made by a British aircraft during 1975. The leasing of aircraft has not been confined to the 707 fleet - during March 1975 two of British Midland's Heralds left Castle Donington on lease to British Island Airways (G-ATIG) and Air Anglia (G-ASVO). On January 31 1975, two of the Viscounts (G-AZLR and G-AZLT) left Castle Donington to start flying scheduled services on behalf of Cyprus Airways from an airstrip at Larnaca. A third Viscount later joined these two aircraft, but by October Cyprus Airways had replaced two of these aircraft with its own Douglas DC9s. British Midland's leasing division can, therefore, be said to be a great success and on August 31 1975 the company announced further contracts totalling £1.5 million worth of business for leases of aircraft to Air Anglia, Cyprus Airways and Malaysian Airlines. Most of these leases involve the operation of services for foreign airlines, and British Midland plays a major role in providing a valuable 'invisible' export for this country.

While the airline was busy promoting its leasing organisation, British Midland did not ignore the development of its scheduled services, and when British Caledonian suspended its Gatwick to Belfast service on October 31 1974, British Midland took over this service. Using Viscounts, the airline opened a thrice daily service over this route on November 1 1974, and after one year of operations a total of 58,000 passengers had been carried on the service. The airline's Viscounts, though, have to compete with British Airways' Tridents and Tristars, but with the introduction of Tristars British Airways did reduce the frequency of its services, and British Midland now runs its Belfast service at a slight profit. Although this service is flown in competition with British Airways, on November 1 1975 Herald G-BAVX opened a thrice daily Birmingham to Heathrow service in co-operation with British Airways who operated a daily service over this route.

British Midland Airways is thus now one of Britain's major independent airlines flying most of its aircraft totally on scheduled services. For the future the airline hopes to open further scheduled services to Europe and to resume inclusive tour charter operations when further jet aircraft are obtained.

Fleet:

Miles M.57 Aerovan 4	G-AJZN	c/n	6420	dd. 6.48,	sold 6.51.
DH89A Dragon Rapide	G-AEAL	c/n	6325	dd. 4.53,	sold 3.56.
	G-AIUK	c/n	6640	dd. 6.48,	sold 2.55.
	G-AIUL	c/n	6837	dd. 6.54,	sold 3.57.
	G-AKME	c/n	6767	dd. 5.50,	written off Lympne 30.6.50, operated by Wolverhampton Aviation.
	G-AKOV	c/n	6612	dd. 8.50,	sold 3.55, operated by Wolverhampton Aviation
Percival Proctor 5	G-AHTE	c/n	Ae.58	dd. 6.55,	sold 9.58.
Douglas C 47A Dakota	G-AGJV	c/n	12195	dd. 3.61,	sold 7.69, 'Millers Dale'.
	G-AKJH	c/n	13164	dd. 5.61,	sold 4.65, 'Monsal Dale'.
Douglas C47B Dakota	G-AMSW	c/n	32919	dd. 4.59,	written off Pyrenees 7.10.61, 'Fern Dale'.
	G-AMSX	c/n	33196	dd. 4.59,	sold 1.66, 'Peak Dale'.
	G-ANTD	c/n	26414	dd. 4.55,	sold 10.68, 'Dove Dale'.
Douglas C47 Dakota	G-AOFZ	c/n	9131	dd. 2.60,	sold 3.66, 'High Dale'.
Douglas C47B Dakota	G-AOGZ	c/n	33282	dd. 6.56,	sold 4.67, 'Darley Dale', later 'Denby Dale'.
	G-APBC	c/n	27121	dd. 5.58,	sold 7.68, 'Derwent Dale'.
Miles M.60 Marathon	G-AMEW	c/n	118	dd. 7.58,	wfu. 10.60.
	G-AMGW	c/n	127	dd.10.55,	wfu. 12.60, 'Millers Dale'.
	G-AMHR	c/n	129	dd.10.55,	wfu. 12.60, 'Monsal Dale'.
Avro 652A Anson 1	G-AIPA	ex.	EF866	dd. 2.58,	wfu. 4.60.
	G-AMDA	ex.	N4877	dd. 7.55,	sold 12.62.
Airspeed AS65 Consul	G-AIKT	c/n	4356	dd. 8.58,	wfu. 3.59, never entered service.
Canadair C4 Argonaut	G-ALHG	c/n	153	dd.10.61,	written off Stockport 4.6.67.
	G-ALHS	c/n	164	dd.10.61,	wfu. 4.68.
	G-ALHY	c/n	170	dd.10.61,	wfu. 6.68.
Handley Page Herald 100	G-APWA	c/n	149	lsd. 4.66,	ret. 9.66, leased from Handley Page.
Handley Page Herald 207	G-ATHE	c/n	165	lsd. 8.65,	ret. 10.65, leased from Handley Page.
Handley Page Herald 211	G-ASKK	c/n	161	dd. 2.65,	sold 12.66.
Handley Page Herald 214	G-ASVO	c/n	185	dd. 3.73	
	G-ATIG	c/n	177	dd. 4.73	
	G-BAVX	c/n	194	dd. 4.73	
DH104 Dove 1B	G-AROG	c/n	04207	dd. 5.63,	sold 3.66.
BAC 1-11 523 FJ	G-AXLL	c/n	193	dd. 1.70,	sold 3.73.
	G-AXLM	c/n	199	dd. 3.70,	sold 4.74.
	G-AXLN	c/n	211	dd. 3.70,	sold 9.73.
Vickers 702 Viscount	G-APPX	c/n	073	lsd. 4.69,	ret. 10.69, leased from Field Aircraft Svs.
Vickers 736 Viscount	G-AODG	c/n	077	dd. 1.67,	written off Castle Donington 20.2.69.
Vickers 755 Viscount	G-AOCB	c/n	092	dd. 2.69,	wfu. 4.70.
	G-AOCC	c/n	093	dd. 2.69,	wfu. 4.69.
Vickers 760 Viscount	G-AWCV	c/n	186	dd. 4.68,	wfu. 4.70.
Vickers 785 Viscount	G-AWGV	c/n	116	dd. 5.68,	wfu. 4.70.
Vickers 813 Viscount	G-AZLP	c/n	346	dd. 1.72	
	G-AZLR	c/n	347	dd. 1.72	
	G-AZLS	c/n	348	dd. 2.72	
	G-AZLT	c/n	349	dd. 2.72	
	G-AZNA	c/n	350	dd. 3.72	
	G-AZNB	c/n	351	dd. 3.72	
	G-AZNC	c/n	352	dd. 3.72	
Vickers 814 Viscount	G-AWXI	c/n	339	dd. 1.69,	written off after fire on take off at Heathrow 22.1.70.
	G-BAPD	c/n	340	dd. 2.73,	sold 9.75.
	G-BAPE	c/n	341	dd. 2.73	
	G-BAPF	c/n	338	dd. 2.73,	sold 10.75.

Vickers 814 Viscount	G-BAPG	c/n	344	dd. 2.73
Vickers 815 Viscount	G-AVJA	c/n	336	dd. 6.67, written off Ringway 20.3.69.
	G-AVJB	c/n	375	dd. 7.67
Vickers 831 Viscount	G-APND	c/n	402	dd. 1.69, sold 11.73.
	G-APNE	c/n	403	dd. 4.67, sold 9.72.
	G-ASED	c/n	419	dd. 2.67, sold 5.72.
Vickers 833 Viscount	G-APTD	c/n	426	lsd. 4.69, ret. 2.70, leased from British United Airways
Boeing 707-321	G-AYBJ	c/n	17597	dd. 4.70
	G-AYVE	c/n	18083	dd. 4.71
	G-AYVG	c/n	17598	lsd. 2.75, leased from Polafern.
	G-AYXR	c/n	17608	lsd. 9.75, leased from Polafern.
	G-AZWA	c/n	17605	lsd. 4.75, leased from Polafern.
	G-BAEL	c/n	17602	lsd. 8.75, leased from Polafern.

BRITISH NEDERLAND

British Nederland Air Services Ltd.

Bases: Tollerton Airport, later Bovingdon Airport.

British Nederland Air Services was an air charter operator registered on June 4 1947 by Mr.J.A.Theones and Mr.A.H.Bryant. Operations started later in June with two Miles Aerovans flying ad-hoc passenger and freight charter work from Tollerton and Blackbushe Airports. However, both Aerovans were withdrawn from service in November 1947, and operations were temporarily suspended while larger aircraft were obtained to allow the airline to resume charter flying. In April 1948 two Douglas Dakotas entered service with British Nederland Air Services. Flying from Bovingdon, these aircraft were initially used for importing fresh fruit from foreign countries to London during April and May. The company was associated with a similar airline operating in Holland named Aero Holland NV of the Hague. This company also operated two Dakotas which were used on ad-hoc passenger and freight charter flights within Europe, and these aircraft were sometimes interchanged with British Nederland's aircraft.

By June 1948 British Nederland's name was becoming well known as a charter operator, and at the end of the month the Dakotas carried a party of sixty three Boy Scouts from Newcastle to Kevic Airport, Kristiansand with a stop at Edinburgh Airport en-route to clear customs. Also during June, two trips were made from Holland to Manchester carrying bilberries, and a Dakota also carried a load of racing pigeons to Guernsey. In July British Nederland transported the sixty three Boy Scouts from Kristiansand back to Newcastle, carried two consignments of prepared animal skins from London to Paris, and also made further flights to Guernsey carrying pigeons. During August one of the Dakotas carried passengers to Paris late in the month and on its return flight the aircraft stopped off at Caen to collect a two and a half ton load of cheese which was then flown back to the United Kingdom. On August 22 a Dakota carried newspapers from Bovingdon to the Channel Islands, and on its return flight the aircraft brought back a load of passengers, while on the following day a ship's crew was flown to Helsinki. British Nederland did make one brief appearance on the Berlin Airlift when, on September 21 1948, Dakota G-AJZX started flying sorties from Lubeck as a freighter aircraft carrying supplies into Berlin. On October 5 this aircraft, like all others based at Lubeck, transferred its operating base to Hamburg, and operations continued from here until November 14 1948 when the Dakota was taken off the airlift. During its two months on the airlift, British Nederland's Dakota operated a total of seventy six sorties during two hundred and thirty flying hours - British Nederland's part in the airlift now over, the Dakota returned to Bovingdon to rejoin the company's other Dakota. British Nederland's two Dakotas continued to undertake a wide variety of ad-hoc charters, and in November the Streatham ice hockey team was flown to the International Ice Hockey Championships in Paris, while another charter later in the month involved the carriage of the musical instruments belonging to the Berlin Philharmonic Orchestra from Bovingdon to Berlin. The orchestra's musicians, though, found it more enjoyable to sail from London to Hamburg, but on arrival in Hamburg the one hundred and three musicians were greeted by a British Nederland Dakota which then proceeded to transport them to Berlin. The last three weeks of 1948 saw the carriage of a consignment of one and a half tons of furs from Bovingdon to Le Bourget, and a load of unexposed cinema film from Bovingdon to Vienna. On the return flight from Vienna, the Dakota brought back some film studio equipment. Another Dakota transported a valuable cargo of antique furniture from London to Paris on December 15, and by the end of the year British Nederland's two Dakotas were some of the hardest worked Dakotas in the country.

While the Dakotas had been busy in 1948 undertaking a wide variety of charter work, British Nederland's Aerovan had also made several welcome appearances at airports while engaged on charter work. These charters included the carriage of racing pigeons from London to Guernsey in June, and of animal skins from London to Paris, and pigeons from London to Twenthe in July. The Aerovan was mainly used for ad-hoc freight charter work, and most of its cargoes were loads of pigeons, furniture, textiles and various types of office and film equipment, but this aircraft was later sold to Aviation Traders at Southend.

The New Year brought with it the first of several contracts from Lep Air Services for the import of textiles from towns on the Continent to the British Isles. These textile contracts were to provide much of British Nederland's work over the next two years, with the first flight taking place on January 6 when a load of textiles was carried from Lille to Bovingdon. This first contract called for the operation of about twenty flights over a period of five weeks. On January 10 one of the Dakotas left Bovingdon with two ship's cylinder coves on board bound for the town of Cochin in India - this flight being one of British Nederland's first long distance charters. The textile flights from Lille to Bovingdon continued during February, and on February 5 one of the airline's Dakotas carried the Bristol United football team from Bristol to Cognac, while a week later the airline flew a ship's crew from Malmo to West Hartlepool and a ship's cylinder liner from Hamburg to Belfast. Also during February, the airline announced the formation of another associate company - registered in South Africa as Air Equator, this charter company started air charter operations with a Beechcraft. On February 18 1949 two tons of plastic sheeting and plastic false teeth were flown in one of the Dakotas from London to Le Bourget. By this time, British Nederland had completed their first contract with Lep Air Services, but with the award of a second contract for a further twenty flights, British Nederland resumed these services between Lille and Bovingdon on March 7 1949. On the same day the other Dakota operated a B.E.A. flight to Prague and Vienna with mixed cargo, while on March 12 two and a half tons of penicillin was flown from London to Milan. On the return trip, the Dakota brought back a three and a half ton load of textiles to Bovingdon.

With Ministry restrictions coming into force on March 31 1949 concerning the import of foreign cloth, British Nederland Air Services increased the number of textile flights operated during March, and during the third week of March alone eleven flights were made from Lille, but services were disrupted at the end of the month by fog. Even with this increase in business, one of the Dakotas still found time to carry a load of passengers from Southend to Hamburg. With the spring approaching, the requirement for the import of foreign fresh fruit and vegetables had now presented itself, but British Nederland was still busy with its textile flights. However, its associate company, Aero Holland, made the first of many flights from the Netherlands to Bovingdon on March 29 carrying a cargo of lettuce for the London markets. During April the company also made other flights carrying lettuces and cucumbers from Holland to Liverpool and Manchester. One of British Nederland's Dakotas left Liverpool on March 29 with a party of 21 wine merchants who wished to sample the vineyards in the Cognac area of France. After three enjoyable days in the area, the party were returned to the sobering surroundings of Liverpool, again aboard a British Nederland Dakota. During March other parties of passengers were taken to the Grand National, while a band of Welshmen were taken from Cardiff to Paris to watch

the France v Wales rugby match.
During April, British Nederland continued its Lille to Bovingdon textile flights, and during the month the fiftieth flight over this route was flown. However, although one Dakota was busily engaged on this work, the other aircraft operated more than a dozen flights during the month flying lettuces, cucumbers and radishes from Amsterdam to Bovingdon. Other ad-hoc charters completed during April included the uplift of almost three tons of machinery from Basle to Manchester on April 14, and the carriage of two tons of films from Northolt to Hamburg on April 23. British Nederland also visited Hamburg on two other occasions during April carrying replacement crews from Southend for airlines still engaged in flying sorties on the Berlin Airlift. This activity became a regular part of British Nederland's operations during the spring and summer of 1949 with many flights taking place between Southend and Hamburg. May saw a reduction in the number of textile flights from Lille following further Government restrictions on the import of foreign cloth, but one Dakota left Bovingdon on May 8 carrying eight passengers and over half a ton of cargo for Lydda. On May 14 ten speedway riders and their motorcycles were flown by Dakota to Antwerp for a race meeting.

By the end of May the vegetable produce flights from Amsterdam had fallen off, and so the Dakotas were switched to importing strawberries from Brest and Valance to Bovingdon during late and June 1949. However, if the regular services of British Nederland had fallen off, the ad-hoc charters had not, as during May and June the Dakotas flew a very wide variety of unusual charters. The Hendon Football Team was flown to Le Bourget on May 21, the Ministry of Civil Aviation's exhibits and representatives at the 1949 Paris Air Show were flown back to London on May 22, three tons of newspapers were airlifted from Northolt to Newcastle due to a rail strike on May 29, two tons of films were flown to Vienna on May 21, and two loads of racing pigeons were carried from Gatwick to Brussels on June 3. During June several more pigeon charters were flown to Brussels, several textile flights were made from Lille and many flights were made to Hamburg and Schleswig carrying spare parts and crews for aircraft operating on the Berlin Airlift. These charters were mainly flown on behalf of Bond Air Services. The newspaper flights to Newcastle and Glasgow also continued during June, operated on behalf of a national newspaper, and a shortage of aircraft during the month led to British Nederland sub-contracting some of its work to Blue-Line Airways and Kearsley Airways. At the end of the month, British Nederland's Dakotas flew three services from Bovingdon to Dublin carrying fruit, carried a party of Boy Scouts from Bovingdon to Norrkoping, and transported a load of ship's spares from Southend to Cherbourg.

During the first week of July, British Nederland's Dakotas carried two parties to the Royal Show at Shrewsbury, and a load of passengers were taken to Stockholm. On the return flight from Stockholm the aircraft brought back twenty two Boy Scouts to this country. Other flights carried out during July included flights to Luxembourg with racing pigeons, and several support flights for airlines operating on the Berlin Airlift. Four fruit carrying flights were made from Le Touquet to Lympne during July, while a party of Boy Scouts was flown from Manchester to Copenhagen, and a group of passengers and their bicycles were carried from Northolt to Paris. Several other flights carrying Boy Scouts were made during July and August with visits being made to several Scandinavian airports - most scouts flew from Manchester to Antwerp, Copenhagen and Norrkoping, from Newcastle to Stavanger and from London to Amsterdam, Copenhagen, Helsinki and Oslo. During July and August British Nederland operated many flights between Le Touquet and Lympne carrying fruit, and similar flights were also made from Spain and other French airports to Bovingdon carrying peaches and apricots. However, so much fruit was imported that markets at Covent Garden could not cope with it, and British Nederland had to fly the fruit onto Dublin before disposing of it. Other ad-hoc charters at this time included the carriage of several more loads of antique furniture to Le Bourget, and other cargo flights were made to Amsterdam, Belfast and Gibraltar mainly carrying ship's spares. The textile services from Lille had died for most of the summer, but on August 8 one of the Dakotas resumed this operation on behalf of Lep Air Services. The ending of the civil participation in the Berlin Airlift during August meant an increase in the number of flights to Hamburg and Schleswig to collect airlift personnel.

September saw British Nederland operating passenger charter flights to Helsinki with students, and to Lisbon with a dance band and its instruments. One notable flight during the month took place towards the end of the month when, after an urgent telephone call, one of the Dakotas flew a five hundredweight iron lung, a doctor and a nurse from Blackbushe to Madrid. The weather for most of the flight was bad, but shortly before one o'clock in the morning the Dakota touched down safely at Muntala Airport, Madrid where a special flare path had been laid for the aircraft. The iron lung was unloaded from the aircraft just under nine hours after the urgent telephone call. At the end of September passenger charters were operated between Birmingham and Ostend, and between Birmingham and Amsterdam, while a party of wine connoisseurs was flown to Cognac and Paris. The textile flights from Lille to Bovingdon continued during October, and many flights were made to Amsterdam during the month. Apart from the carriage of ship's crews to Amsterdam, British Nederland also flew both of the Dakotas from Bovingdon to Amsterdam on October 25 carrying the Metropolitan Ballet Company. During November numerous flights were made from Amsterdam to Bovingdon carrying passengers or grapes, but one unusual charter undertaken during the month was the carriage of a load of mixed animals to Nairobi. This flight left the British Isles on November 21, and aboard the Dakota were calves, dogs, cats and chickens belonging to the British Livestock Export Company Ltd. As 1949 drew to a close, business fell off but British Nederland were fortunate in obtaining a contract to fly three services from Bovingdon to Vienna on behalf of the British Livestock Export Company Ltd. The first of these flights took place on November 30 with a cargo of pigs aboard, and the remaining two flights were undertaken early in December.

Thus, another year of successful operations had drawn to a close, and British Nederland's two aircraft were still some of the hardest worked Dakotas in Europe. The opening weeks of 1950, though, were very quiet for the airline and, apart from the transportation of some tabulating machinery to Belfast and a few passenger and freight charters to Amsterdam and Paris, British Nederland's aircraft were unable to find much work. During February one of the Dakotas carried a David Brown tractor, a plough and a harrow from Bovingdon to Paris. In March 1950, business began to pick up again and although one of the Dakotas went away for a C of A inspection, the other aircraft continued operating charter flights. Occasional trips were still made to Lille to collect textiles,and other charters completed during March included the carriage of Welsh rugby supporters on two trips from Cardiff to Belfast, and a party of representatives of the Birmingham Chamber of Commerce were flown to Bristol to view the Brabazon. By the end of March the number of textile flights had once again increased, and two flights were made to Brussels at the end of the month - one carrying a group of athletes and the other with a consignment of X-ray equipment. One of the Dakotas carried an aero engine to Tripoli on March 28, while the other aircraft carried theatre scenery and props to Madrid on the same day. At the end of March British Nederland's Dutch associate, Aero Holland, was forced to stop operating after the Dutch Government had placed further restrictions on the operation of charter flights by private companies.

During the spring of 1950 British Nederland operated a series of passenger charter flights to Amsterdam from Birmingham and Bovingdon. During April one of the Dakotas flew three tons of machinery from Bovingdon to Stockholm, while a passenger charter flight was made from Croydon to Le Bourget. One of the Dakotas flew a load of film equipment to Nice on April 20, while a cargo consisting of ten pigs, seven cows and two calves were flown to Milan and Malta on March 19. During April, British Nederland also operated several flights from Amsterdam to Bovingdon carrying Dutch vegetable produce.

In April 1950 it was announced that British Nederland Air Services had been awarded a contract by the British Government to run a leave service for staff of the British Administration in Tripolitania. In order to operate this service, British Nederland acquired a Vickers Viking, and this aircraft left on the first service on May 16 1950. This service was subsequently flown at the rate of approximately one service each month and, by the time the third flight left Bovingdon for Castel Benito on July 18 1950, the twenty one seat Viking was operating at a one hundred per cent load factor. With only twenty one seats fitted to this aircraft, British Nederland was able to offer its passengers a spacious cabin and allow them to carry sixty six pounds of luggage each. During May several more textile flights were made from Lille, but the imported cloth was now flown to Southend where Lep Air Services had opened an office to deal with this work. May also saw the continuation of flights between Amsterdam and Bovingdon carrying cucumbers, lettuces and radishes. From Birmingham, British Nederland flew a party of factory workers to Dinard on May 6, and another party of fifty to Amsterdam on May 13. Other ad-hoc charters flown during May included the uplift of film equipment from Nice to Paris and Northolt on May 11, the carriage of a load of oil pumps from Bovingdon to Antwerp on May 10, and the transport of two loads of passengers to Paris on May 26 and 27.

On June 2 1950 one of the Dakotas flew ninety baskets of pigeons from Gatwick to Nantes, and on the occasion of its third birthday British Nederland flew a party of American tourists from Luxembourg to Geneva. The remainder of June saw no particularly startling charters being operated by British Nederland unless the operation of further pigeon flights to Cannes comes into this category. July was rather busier and apart from pigeon flights to Caen and Nantes, the Dakotas flew a ship's crew from London to Bremen, Scout troops from London to Aarhus, Copenhagen, Helsinki, Oslo and Stockholm, NAAFI staff to Cairo, and a mixed load of animals to Lisbon. Passenger charters between Amsterdam and Bovingdon continued into August, and these flights were often operated to collect passengers from Braathens' aircraft at Amsterdam, and then to fly them on to London. The movement of Scout troops during August was a lucrative source of income for British Nederland and its Dakotas operated flights to Amsterdam, Copenhagen, Helsinki and Innsbruck carrying Boy Scouts. Several tourist flights carrying holidaymakers were also flown during August, and popular destinations included Oslo, Salzburg, Stavanger and Stockholm.

On September 13 British Nederland's entire fleet of aircraft - the Viking and two Dakotas - flew seventy five farmers from Cambridge to Liverpool. Six days later one of the Dakotas left Bovingdon for Tripoli and Lebanon carrying cargo on behalf of the Iraq Petroleum Company, while on September 25 another aircraft left Blackbushe for Cairo carrying NAAFI personnel. By October, though, the work was declining fast - the textile and vegetable produce flights had come to an end, and although one of the Dakotas carried members of the Halle Orchestra to Lisbon on October 6, few other international charter flights were made during October.

With the high prices then being paid for good second hand Dakotas, British Nederland sold one of its aircraft in November 1950, and charter operations continued with the Viking and one Dakota. On November 28 the airline carried NAAFI staff to Cairo, and a cargo of cows, pigs and chickens to Milan. However, by December work had become scarce, and with insufficient work to fully occupy these two aircraft, British Nederland Air Services ceased all operations on December 31 1950.

Fleet:

Miles M57 Aerovan	G-AISE	c/n	6395	dd. 6.47,	wfu. 11.47.
	G-AISI	c/n	6397	dd. 5.48,	sold .49.
	G-AJTK	c/n	6416	dd. 6.47,	wfu. 11.47.
Douglas C47A Dakota	G-AJZD	c/n	12333	dd. 3.48,	sold 11.50.
Douglas C47 Dakota	G-AJZX	c/n	9051	dd. 4.48,	ISOC.
Vickers 498 Viking	G-AHOP	c/n	117	dd. 1.50,	ISOC.
	G-AHOV	c/n	123	dd. 7.50,	wfu. 10.50, never entered service.

BRITISH UNITED

British United Airways Ltd.
Airwork Ltd.
Air Charter Ltd.
Hunting-Clan Air Transport Ltd.
Morton Air Services Ltd.
Olley Air Service Ltd.
Silver City Airways Ltd.
Transair Ltd.

Base: London (Gatwick) Airport.

Proposals were first put forward for the merger of Airwork and Hunting-Clan in March 1960, but it was not until June 27 1960 that the final plans were announced by Airwork and Hunting-Clan for the merger of the two airlines, and their subsidiaries, into one unit. This merger came into effect on July 1 1960 and the new group comprised Airwork, Air Charter, Airwork Helicopters, Bristow Helicopters, Hunting-Clan, Morton Air Services, Olley Air Service and Transair. The new airline was named British United Airways and its operations were broken down into three separate divisions - an Air Transport Division, a Helicopter Division and an Overseas Contracts Division. The chairman of British United was Mr.M.D.N.Wyatt, CBE, while Mr.Freddie Laker became managing director - other directors included Messrs. R.L.Cumming, R.C.Benbow, A.E.Bristow, T.W.Morton, B.R.Seton-Winton, J.A.Thompson and the Hon.A.Cayzer. At the time of its formation, British United was operating a fleet of Viscounts, Britannias, Dakotas, DC4s and DC6s. The Bristol Freighters operated on cross-Channel vehicle ferry services from Southend continued to operate under the name of Channel Air Bridge, although they, too, were part of British United Airways. British United flew scheduled passenger services to East, Central and West Africa under the title of 'Britannia Thruliner' and 'Skycoach' services, while other schedules were flown from Gatwick to Gibraltar, Rotterdam, Le Touquet, Jersey and Guernsey. A Blue Nile service was flown on behalf of Sudan Airways to Rome, Athens, Cairo, Khartoum, Beirut, Aden and Asmara using Viscounts, and Africargo services were flown from Heathrow via Germany and Malta to East, Central and South Africa. A large amount of -scheduled work was also carried out, and this mainly involved the operation of government freighting and trooping contracts. The trooping services were mainly flown fron Stansted with Britannias, and destinations served included Hong Kong, Singapore, Nairobi, Aden, Cyprus, Malta and several bases within Germany including Gutersloh and Wildenrath.

In October 1960 British United's Viscount 810s opened a Skycoach service from London to Africa in partnership with B.O.A.C., East African Airways and Central African Airways. These services offered a much lower air fare than was charged for a normal scheduled service, but the passengers still travelled in the comfort of a sixty two seat Viscount. During the first twelve months of these services, British United operated forty six flights to East and Central Africa - in addition to their own services, British United's Viscounts also flew thirteen services to East Africa on behalf of East African Airways, and another thirteen to Central Africa on behalf of Central African Airways. The Viscounts proved to be popular on these services, but with the award of several European trooping contracts, British United decided to switch these aircraft to these trooping flights. Thus, on October 2 1961 the Viscounts were replaced by eighty seven seat Britannias, and the Viscounts were transferred to operating trooping flights from Gatwick to Royal Air Force bases in West Germany. On October 2 British United Airways became responsible for all movements of servicemen and their families from the United Kingdom to Germany, as the authorities had ceased to use the surface route via the Harwich to Hook sea ferry to West Germany. From this date the Viscounts, and also later the DC6s, carried eleven thousand servicemen each month from Gatwick and Manchester to Germany including Dusseldorf, Frankfurt, Gutersloh, Hanover and Wildenrath. The first service was flown by Viscount G-APNE from Gatwick to Hanover on October 2 1961, and later the same day services were also flown to Dusseldorf and Gutersloh. Following this, on October 7, British United inaugurated a once weekly service from Gatwick to Tenerife.

British United Airways had, therefore, become Britain's first really major independent airline, and this was further confirmed when on May 9 1961 the airline became the first operator in the world to order the new BAC 1-11. Ten of these short haul jet airliners were ordered to replace the Viscounts, and to help British United to set up new scheduled services to Europe. In January 1962, the British Aviation Services Group was bought out by British United, and Silver City Airways thus became part of the British United Group. Silver City were operating a wide variety of scheduled services with Dakotas, Bristol Freighters, Hermes and three newly acquired Viscounts. With the acquisition of British Aviation Services, a new holding company for the two organisations was formed named Air Holdings. The Air Holdings' board consisted of ten people, of which eight were from British United Airways - these ten directors were Mr.Myles Wyatt (Chairman), Sir Nicholas Cayzer (Deputy Chairman), Sir Donald Anderson, Mr.P.B.Guiness, Mr.F.A.Laker, Sir Brian Mountain, Mr.G.N.A.Murrant, Lord Poole, Mr.G.H.Trott and Mr.R.L.Cumming. The main shareholders in Air Holdings at this time were the P and O Shipping Company; the Eagle Star Insurance Group; Cable and Wireless Holdings; the Blue Star Line; the British and Commonwealth Shipping Company; Clan Line Steamers; Broadminster Nominees; Furness, Withy and Company; Guiness, Mahon Nominees; and Hunting Aviation Management. During 1962 a reorganisation of all the airlines within the group took place,resulting in the formation of British United (Channel Islands) Airways on November 1 1962, and of British United Air Ferries on January 1 1963. Prior to this, on May 20 1962, Jersey Airlines had also become part of the Air Holdings Group, and British United (Channel Islands) Airways now became responsible for the operation of most of the scheduled passenger and cargo services to the Channel Islands and the Isle of Man using a large fleet of Dakotas and Heralds. British United Air Ferries was a combination of Channel Air Bridge and the vehicle ferry division of Silver City Airways, and most of their services were flown from Southend, Lydd and Southampton to cross-Channel destinations using Bristol Freighters and Carvairs. Meanwhile, British United Airways now operated all the trooping services, the Africargo flights and several important scheduled services, in addition to charter services.

The Africargo service was flown once weekly from Heathrow Airport by one of British United's DC6s. However, between September 1 1962 and August 31 1963, there was an increase of about twenty six per cent in tonnage carried on this service, resulting in the addition of a second weekly DC6 all-freight service to East Africa in October 1963. British United's fleet of Viscounts was also used on inclusive tour charter flights from Gatwick, and during 1963 these services were flown to Barcelona, Basle, Deauville, Lisbon, Milan, Munich, Nice, Palma, Pisa, Rimini, Tarbes, Toulouse and Valencia. By January 1964 British United Airways' services linked Gatwick with East, West and Central Africa, Barcelona, the Canary Islands, Genoa, Gibraltar, Le Touquet, Malaga, Palma, Rotterdam and Tarbes. A twice weekly Blackpool - Manchester - Bournemouth service was inaugurated on January 1 1964 using Dakotas, and then on March 12 a weekly Viscount service was opened between Gatwick and Seville in Spain. During 1964 and early 1965 the BAC 1-11 was operated on proving flights on British United's network to Europe, the Mediterranean and West Africa, and on April 5 1965 the aircraft was granted its full C of A, followed on April 9 by the operation of the world's first commercial 1-11 service from Gatwick to Genoa by G-ASJJ. Jet services to Rotterdam soon followed, and then the aircraft entered service on British United's routes to West Africa.

In addition to ordering 1-11s, British United had also ordered two standard VC10s from Vickers for use on its long distance scheduled services, and on trooping flights. The Britannias were still operating their trooping flights from Stansted, but the last Britannia trooping flight left Stansted on the night of October 1 1964. Subsequently, all trooping flights were flown from Gatwick, and the first VC10 started flying these services on October 1. The first flights served Aden, Bahrein and Nairobi, and the aircraft were operated with 129 rearward facing seats, and there was also provision for the carriage of up to 4,000 lb of freight on each flight. The VC10s were also used on a twice weekly scheduled service between Gatwick and Santiago in Chile - a route taken over by British United from B.O.A.C. By the end of 1968, British United Airways was able to announce that it was now making a modest profit on the scheduled services to South America after almost four years of operation. The service to Santiago in Chile, opened in the spring of 1965, was now flown via Lisbon or Madrid, Las Palmas, Rio de Janeiro, Sao Paulo, Montevideo and Buenos Aires. The VC10 also soon replaced Britannias on the services from Gatwick to Salisbury (Southern Rhodesia) and Accra, and the passenger appeal of these aircraft and the success of their operations prompted British United to order a third aircraft.

During 1965 the 1-11s had made many proving flights to airports all over the British Isles - G-ASJF made the first of a series of proving flights into Manchester on March 17, while later in the year other airports were also visited in connection with the introduction, early in 1966, of British United's new 'Interjet' schedules from Gatwick. These services were flown by BAC 1-11s from Gatwick to Belfast, Edinburgh and Glasgow. On December 13 1965 G-ASJI operated a proving flight from Gatwick to Turnhouse on behalf of the Press, followed on January 4 1966 by the inauguration of these services, with 1-11 G-ASJJ flying the first Gatwick to Belfast service. These services immediately proved popular and were flown in competition with B.E.A. who were flying mainly, Vanguards on all these routes, although British Eagle flew 1-11s and Britannias on their Heathrow to Glasgow service. The interjet services were flown twice daily on all routes, and were timed to connect with many of British United's international scheduled services from Gatwick.

By 1968 the route network was very extensive covering South America, Central, South, West and East Africa, and many parts of Europe. Inclusive tour work was also a major activity and tours were flown to many parts of Europe and North Africa on behalf of numerous tour companies including Leroy Tours, Lord Brothers Holidays, Mediterranean Villas, Horizon, Global and Lyons Tours. The 1969 inclusive tour programme flown for Lyons was for 1-11 departures from Gatwick to Alicante, Basle, Bergen, Genoa, Gerona, Ibiza, Milan, Munich, Naples, Palma, Rimini, Tangier and Venice; from Luton to Palma; and from Manchester to Basle, Gerona and Palma. In addition, the Viscounts were also used on flights from Luton to Gerona, and from Gatwick to Gerona, Munich, Rimini and Venice. Between April and June 1969 five of the new 'stretched' series 500 1-11s entered service for use specifically on these inclusive tour services, and by the end of the year British United's passenger total had increased by fifty three per cent, compared with 1968, carrying over 1,400,000 passengers. On non-scheduled operations the increase was fifty eight per cent, while on scheduled services the increase was forty four per cent, with the interjet services to Belfast, Edinburgh and Glasgow carrying 180,000 passengers. This very encouraging year caused British United to order a further three 1-11 series 500s for delivery early in 1970. 1970's inclusive tour charters were even more extensive than those of 1969 and destinations added included Dubrovnik, Faro, Ljubljana, Pula, Tenerife and Zurich, and services were also flown from Birmingham to Ibiza and Palma. Trooping services were still flown to many German Royal Air Force bases from Gatwick, although this activity had become less important. The Africargo services were flown from Heathrow by all-freighter Britannia aircraft from 1966 up until the end of 1968, but with the purchase of a fourth VC10 early in 1969, these services were now flown from Gatwick with VC10s.

During 1970, Air Holdings tried to find a buyer for British United Airways, and although the two State Corporations were approached with a view to a take-over of British United by B.O.A.C., it was decided at a meeting on October 20 1970 that Caledonian Airways would purchase the whole of the ordinary share capital of British United Airways for £6,900,000. This take-over became effective on November 30 1970, and the new airline became known as Caledonian // BUA and later as British Caledonian Airways.

Fleet:

Vickers 708 Viscount	G-ARBY	c/n	010	ISOF	, sold 2.67.
	G-ARER	c/n	012	ISOF	, sold 6.66.
	G-ARGR	c/n	014	ISOF	, sold 9.66.
Vickers 736 Viscount	G-AODG	c/n	077	ISOF	, sold 1.67.
Vickers 831 Viscount	G-APND	c/n	402	ISOF	, sold 1.69.
	G-APNE	c/n	403	ISOF	, sold 4.67.
	G-ASED	c/n	419	dd.12.62	, sold 2.67.
Vickers 833 Viscount	G-APTB	c/n	424	ISOF	, sold 12.69.
	G-APTC	c/n	425	ISOF	, sold 10.69.
	G-APTD	c/n	426	ISOF	, sold 4.69.
Vickers 804 Viscount	G-AOXU	c/n	248	ISOF	, sold 12.62.
	G-AOXV	c/n	249	ISOF	, sold 11.62.
	G-APKG	c/n	395	ISOF	, sold 11.62.
Douglas C54B Skymaster	G-ANYB	c/n	10528	ISOF	, to B.U.A.F. as a Carvair.
Douglas C54A Skymaster	G-AOFW	c/n	10351	ISOF	, to B.U.A.F. as a Carvair.
Douglas DC4	G-AOXK	c/n	42931	ISOF	, sold 7.63.

Douglas C54A Skymaster	G-APID	c/n	10408	lsd. 1.61, ret. 3.61, leased from Trans World Leasing.
Douglas C54B Skymaster	G-APNH	c/n	18333	ISOF , to B.U.A.F. as a Carvair.
Douglas C54A Skymaster	G-AREK	c/n	10365	dd. 8.60, to B.U.A.F. as a Carvair.
Douglas DC6A	G-APNO	c/n	45531	ISOF , sold 12.65.
	G-APNP	c/n	45532	ISOF , sold 12.65.
Douglas DC6B	G-ARXZ	c/n	45326	lsd. 3.62, ret. 9.64, leased from Canadian Pacific.
Bristol 175 Britannia 307	G-ANCD	c/n	12920	ISOF , sold 1.69.
	G-ANCE	c/n	12921	ISOF , sold 2.69.
Bristol 175 Britannia 309	G-ANCH	c/n	12924	dd. 6.65, sold 9.68.
Bristol 175 Britannia 312	G-AOVE	c/n	13236	dd.10.61, sold 6.64.
	G-AOVI	c/n	12926	dd. 9.61, sold 12.64.
Bristol 175 Britannia 313	G-ARWZ	c/n	13233	dd. 2.62, sold 3.65.
	G-ARXA	c/n	13234	dd. 3.62, sold 5.66.
	G-ASFU	c/n	13431	lsd. 3.63, ret. 3.63. Operated for El Al.
	G-ASFV	c/n	13232	lsd. 3.63, ret. 3.63. Operated for El Al.
Bristol 175 Britannia 317	G-APNA	c/n	13425	ISOF , sold 10.67.
	G-APNB	c/n	13426	ISOF , sold 3.69.
Bristol 175 Britannia 319	G-AOVA	c/n	13207	lsd. 2.64, ret. 4.64, leased from Bristol Aircraft.
Cessna 310D	G-ARAC	c/n	39154	rd. 7.60, transferred to Airwork Services, Perth.
Cessna 320A	G-ASDJ	c/n	0032	dd. 2.64, transferred to Airwork Services, Perth.
Handley Page Herald 204	G-ASBP	c/n	163	dd.11.62, sold 10.66, V.I.P. Transport.
BAC 1-11 201AC	G-ASJA	c/n	005	dd.10.65, sold 10.69.
	G-ASJC	c/n	007	dd.11.65, ISOT.
	G-ASJD	c/n	008	dd. 8.65, ISOT.
	G-ASJE	c/n	009	dd. 7.65, ISOT.
	G-ASJF	c/n	010	dd. 5.65, ISOT.
	G-ASJG	c/n	011	dd. 7.65, ISOT.
	G-ASJH	c/n	012	dd. 4.65, ISOT.
	G-ASJI	c/n	013	dd. 4.65, ISOT.
	G-ASJJ	c/n	014	dd. 4.65, written off Milan, Linate 14.1.69.
	G-ASTJ	c/n	085	dd.11.65, ISOT.
BAC 1-11 408EF	G-AWKJ	c/n	128	lsd. 4.69, ret. 10.69, leased from Channel Airways.
BAC 1-11 501EX	G-AWYR	c/n	174	dd. 4.69, ISOT.
	G-AWYS	c/n	175	dd. 4.69, ISOT.
	G-AWYT	c/n	176	dd. 5.69, ISOT.
	G-AWYU	c/n	177	dd. 6.69, ISOT.
	G-AWYV	c/n	178	dd. 6.69, ISOT.
	G-AXJK	c/n	191	dd. 3.70, ISOT.
	G-AXJL	c/n	209	dd. 3.70, ISOT.
	G-AXJM	c/n	214	dd. 3.70, ISOT.
BAC VC10 1109	G-ARTA	c/n	803	dd. 4.69, ISOT.
BAC VC10 1103	G-ASIW	c/n	819	dd. 9.64, ISOT.
	G-ASIX	c/n	820	dd.10.64, ISOT.
	G-ATDJ	c/n	825	dd. 1.65, ISOT.

BRITISH WESTPOINT

British Westpoint Airlines Ltd.
Westpoint Aviation Ltd.
Mayflower Air Services Ltd.
F & J Mann Airways Ltd.

Base: Exeter Airport.

On December 2 1960 Messrs Frank H.J.Mann and Jack F.Mann registered an airline company named F and J Mann Airways with a share capital of one hundred pounds. However, in March 1961 Dakota operations started from Exeter Airport under the title of Westpoint Aviation, and in addition to F.H.Mann and J.F.Mann, Captain D.R.Mann also joined the board of directors. Based at Exeter Airport, Westpoint Aviation operated a variety of ad-hoc passenger and freight charters with the Dakota, and during 1961 the company was awarded a seven year licence for the operation of a scheduled service between Exeter and Newcastle. However, this service was never flown, and during the summer of 1961 the airline turned its attentions to the operation of inclusive tour charter flights. From July onwards, Westpoint's Dakotas became a familiar sight at Gatwick Airport carrying holidaymakers from there to the sunshine of Dieppe, Innsbruck and Nice. A second Dakota was bought in July 1961 and this aircraft made its first appearance at Gatwick on September 24 1961 when it operated an inclusive tour flight from Nice to Gatwick. During the winter of 1961 the two Dakotas increased the number of services flown from Gatwick, and during September and October new inclusive tour services were opened to Jersey, Le Bourget, Lyon, Maastricht, Munich, Perpignan and Tarbes. Le Bourget was the most popular destination served, and on some days two services in each direction were flown between Gatwick and Paris.

Westpoint's charter flying was not only confined to Exeter and Gatwick, on several occasions the Dakotas flew charters to Hurn and Leeds, and Westpoint's Dakotas were often chartered by other airlines to operate their flights - on August 3 1961 G-ALYF operated a Jersey to Hurn service on behalf of Air Safaris. The Dakotas would also often carry out crew training sorties at Gatwick, and on some occasions the aircraft would fly from Exeter to Gatwick, carry out a crew training detail, and then return westwards to Exeter when the detail was completed.

The winter of 1961 was, therefore, a busy one for Westpoint, and for 1962 the airline had planned to open inclusive tour services from Newcastle to Bergen, Jersey and Ostend, but these routes were never opened. However, for the 1962 season Westpoint had three Dakotas in service - again inclusive tour flights were operated from Gatwick, and from Exeter several cross-Channel flights were made to airports in North Western France including Cherbourg, Dinard and Quimper. During 1962 Westpoint applied to the A.T.L.B. for permission to operate a scheduled vehicle ferry service between Exeter and Cherbourg using Bristol Freighters, but this service was not approved, and Westpoint dropped its plans for any further vehicle ferry services. Over the next four years, Westpoint applied for numerous scheduled services, but although many were approved, very few were ever flown.

For the 1963 season, the fleet remained unchanged at three Dakotas, and on April 1 1963 G-AMDE opened Westpoint's first scheduled passenger service when it inaugurated a Newquay to Heathrow, via Exeter, service, which was flown on weekdays only. Although Westpoint continued to operate inclusive tour charters from Gatwick, the number of services was reduced, and the Dakotas flew mainly to Copenhagen, Deauville, Paris and Perpignan. In October 1963 British Westpoint (as it was now known) started negotiations with Mayflower Air Services for the take-over of their services, but the negotiations were not successful, and the ownership of Mayflower passed into the hands of Scillonian Air Services.

Early in 1964 British Westpoint started flying the London to Lille scheduled service on behalf of Air France - the Dakota G-ALYF was used for this service with 'Air France' stickers on the fuselage of the aircraft. Then, in May 1964, at the second attempt, British Westpoint acquired control of Mayflower Air Services. From Mayflower, Westpoint inherited two Rapides and a network of scheduled services to the Scilly Isles - with this take-over, therefore, Westpoint was able to offer air links with the Scillies from Bristol, Cardiff, Exeter, Newquay and Plymouth. In addition, passengers from London who wished to travel to the Scilly Isles could fly from Heathrow to Exeter or Newquay on the Dakota service, and then clamber aboard a Dragon Rapide and be flown sedately to the Scillies. British Westpoint also obtained a Plymouth to London licence as a result of the acquisition of Mayflower, but this service was never opened. During the summer, British Westpoint operated approximately one to two services a week to Cardiff and Bristol from the Scilly Isles - many services, flown by Captains Birch or Miller, were flown via Exeter or Plymouth where further passengers could be collected, but at the height of the season most of the services were flown direct to St.Mary's. In addition to Mayflower's aircraft, British Westpoint also purchased the last three B.E.A. Dragon Rapides and thus kept this grand old lady in the skies of the South West of England for a few more years. During 1964 the Dakotas continued to operate numerous ad-hoc charters, and visits were made to many airports including Cardiff, Dublin, Glasgow, Leeds, Liverpool and Manchester, but the inclusive tour flights from Gatwick had come to an end. One of the Dakotas was also used as a support aircraft for the HS125 during some of its trials abroad, and the absence of this aircraft necessitated the leasing of a Dakota from Aviation Overhauls for the season. 1964 also saw Westpoint applying to the A.T.L.B. for permission to operate further services - these included a London to Sandown service which was approved in July, and an Exeter to Ostend service approved in December, but an Exeter to Birmingham and Leeds services was not approved.

By 1965 the Exeter to London service had become infrequent, but the London to Lille service continued unchanged. During the year, with further expansion in view, British Westpoint received a demonstration of an Avro 748, and also looked into the possibility of acquiring Air France Douglas DC4s, but these plans came to nothing. A thrice weekly Exeter to Birmingham service using Dakotas was approved in February 1965, but like the services to Sandown and Ostend, no flights on these routes ever took place. During the summer months, the Rapides once again operated the scheduled services to the Scilly Isles, and these aircraft were also used for pleasure flying at many of the local air displays in the South West of England during the summer months. In August 1965, British Westpoint took over the newspaper runs from Ringway which had previously been flown by South Coast Air Services. These flights were operated on six nights a week, and using Dakotas the services linked Ringway with Shannon. The other Dakotas, meanwhile, continued flying many ad-hoc passenger and freight charters, and in October 1965 British Westpoint was bought out by Metropolitan Air Movements. This company then became the holding company for British Westpoint, but Westpoint continued to operate under its own name. In November 1965 an Exeter to Rotterdam service was approved, but never operated, and in January 1966 a Heathrow to Gatwick shuttle service using Rapides was approved, but yet again this service was never flown.

The winter of 1965 was a quiet time for British Westpoint, and although the Ringway newspaper services provided a welcome source of income, the airline ran into financial problems early in 1966. This brought any further expansion plans to an abrupt halt, and early in May 1966 British Westpoint Airlines ceased all flying, and was put into liquidation.

Fleet:

DH89A Dragon Rapide	G-AGSH	c/n 6884	dd. 5.64,	ISOC.		
	G-AHKU	c/n 6810	dd. 5.64,	ISOC.		
	G-AIUL	c/n 6837	dd. 5.64,	sold 10.64.		
	G-AJCL	c/n 6722	dd. 5.64,	ISOC.		
	G-ASKO	c/n 6735	dd. 5.64,	sold 5.65.		
DH104 Dove 1	G-AKET	c/n 04056	lsd. .65,	ret. .65.		
DH104 Dove 2	G-ALCU	c/n 04022	lsd. .65,	ret. .65.		
DH104 Dove 5	G-AROI	c/n 04474	dd.10.65,	ISOC.		
DH114 Heron 2E	G-ATCZ	c/n 14076	dd. 4.65,	sold 3.66.		
Douglas C47A Dakota	G-AJHY	c/n 13388	dd. 7.61,	sold 10.64,	'Sir Walter Raleigh'.	
	G-AJHZ	c/n 12421	lsd. 3.65,	ret. 4.65,	leased from South Coast Air Services.	
	G-ALYF	c/n 19350	dd. 3.61,	ISOC	, 'Sir Francis Drake'.	
Douglas C47B Dakota	G-AMDB	c/n 26432	dd. 3.62,	ISOC	, 'Sir Richard Grenville'.	
	G-AMPO	c/n 33186	lsd. 1.64,	ret. 12.64,	leased from Aviation Overhauls.	

There now follows a short appendix containing a number of airline route maps showing scheduled services only. These maps are chosen at random and do not represent a complete picture of any airline's scheduled services networks.

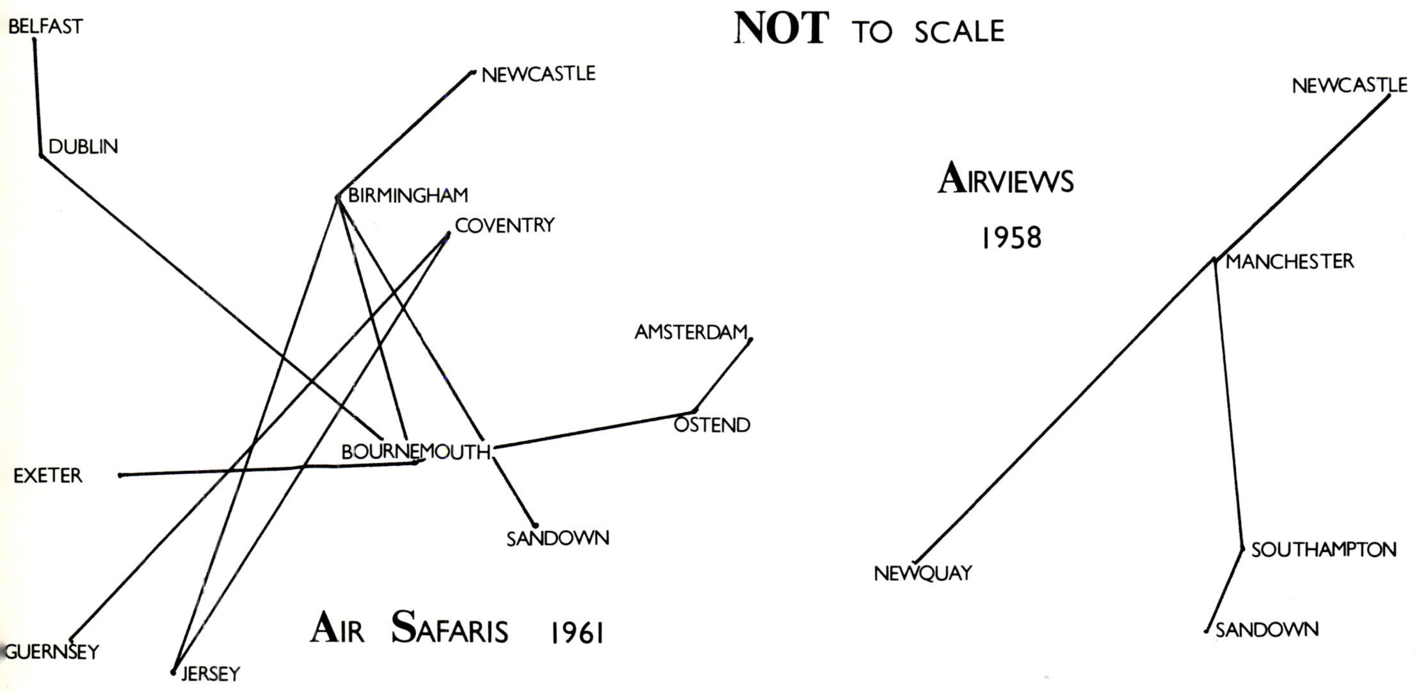

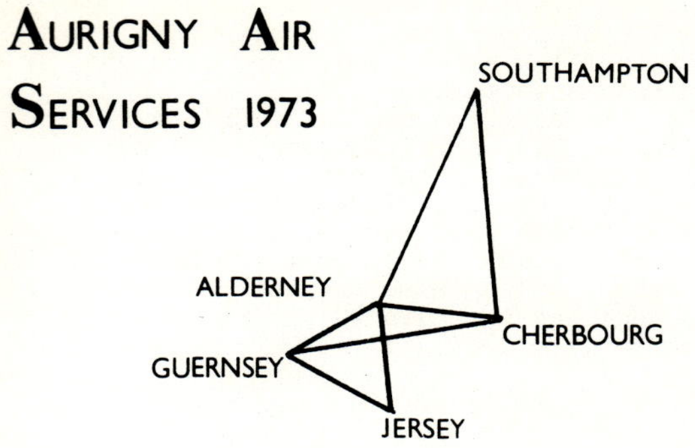
AURIGNY AIR SERVICES 1973
SOUTHAMPTON
ALDERNEY
CHERBOURG
GUERNSEY
JERSEY

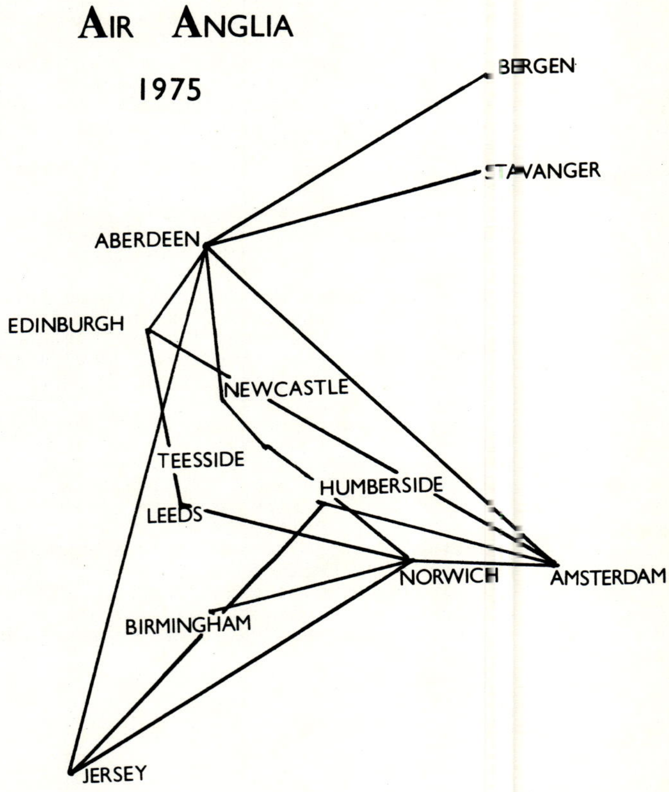
AIR ANGLIA
1975
BERGEN
STAVANGER
ABERDEEN
EDINBURGH
NEWCASTLE
TEESSIDE
HUMBERSIDE
LEEDS
NORWICH
AMSTERDAM
BIRMINGHAM
JERSEY

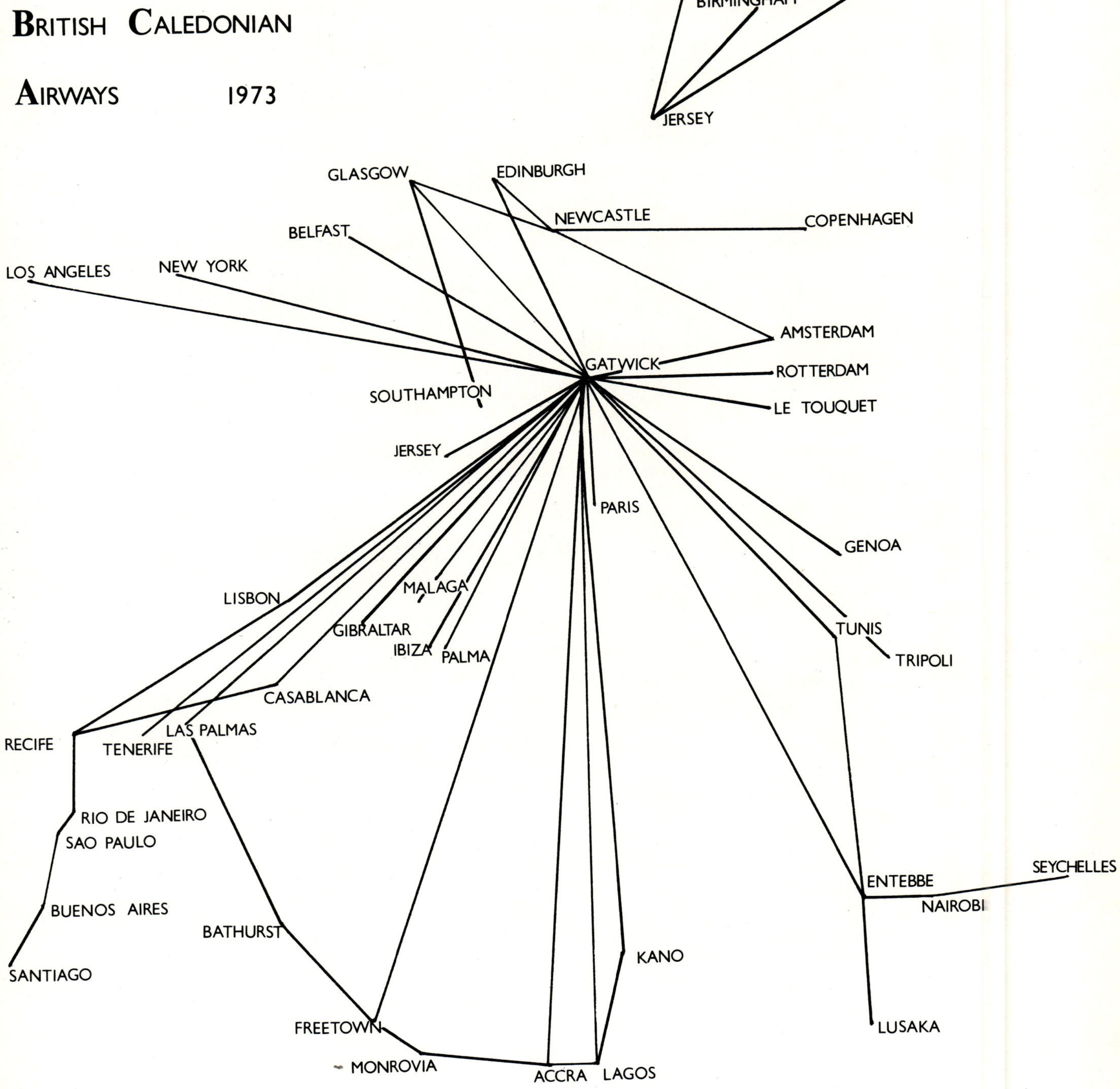
BRITISH CALEDONIAN AIRWAYS 1973
GLASGOW
EDINBURGH
NEWCASTLE
COPENHAGEN
BELFAST
LOS ANGELES
NEW YORK
AMSTERDAM
GATWICK
ROTTERDAM
SOUTHAMPTON
LE TOUQUET
JERSEY
PARIS
GENOA
LISBON
MALAGA
GIBRALTAR
IBIZA
PALMA
TUNIS
TRIPOLI
CASABLANCA
RECIFE
TENERIFE
LAS PALMAS
RIO DE JANEIRO
SAO PAULO
BUENOS AIRES
SANTIAGO
BATHURST
ENTEBBE
SEYCHELLES
NAIROBI
KANO
FREETOWN
MONROVIA
ACCRA
LAGOS
LUSAKA

Eagle Airways

1958

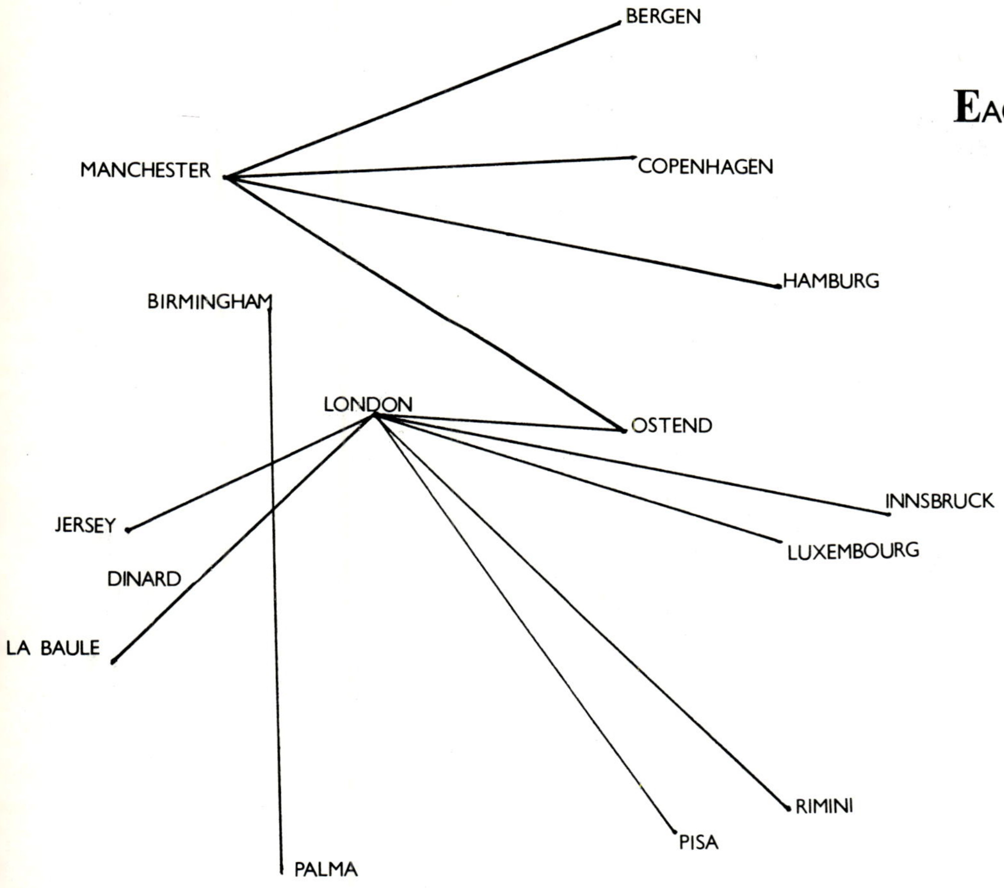

British Eagle International Airlines 1966

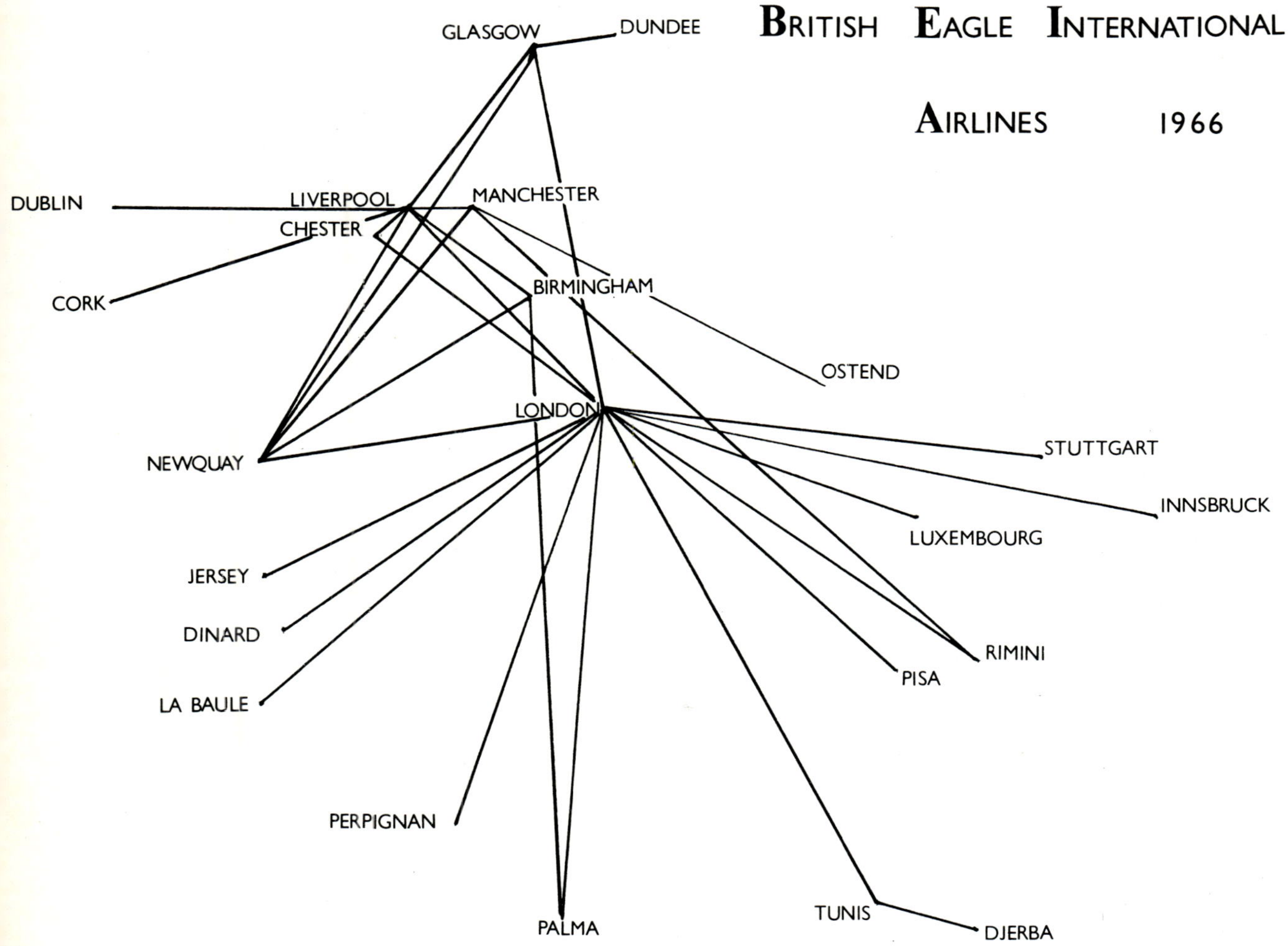

British Island Airways 1972

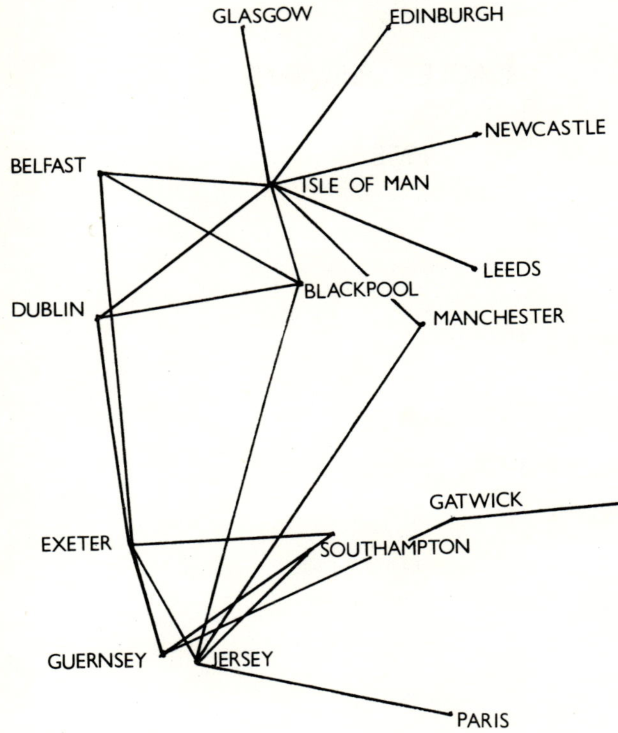

British Midland Airways 1967

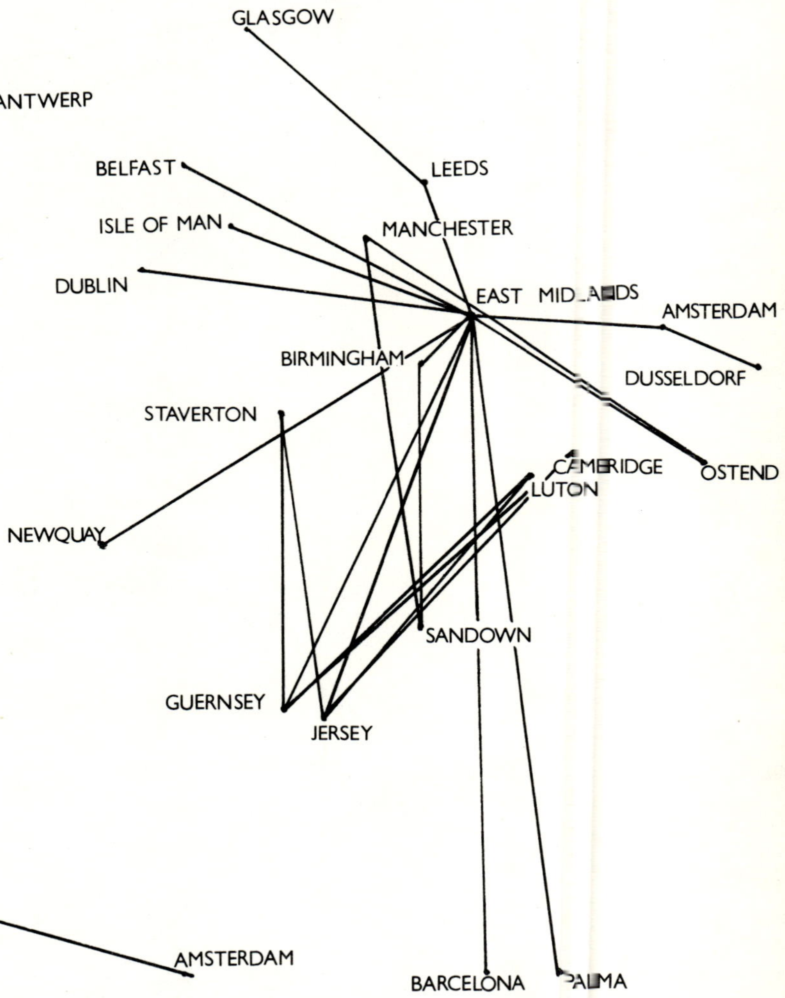

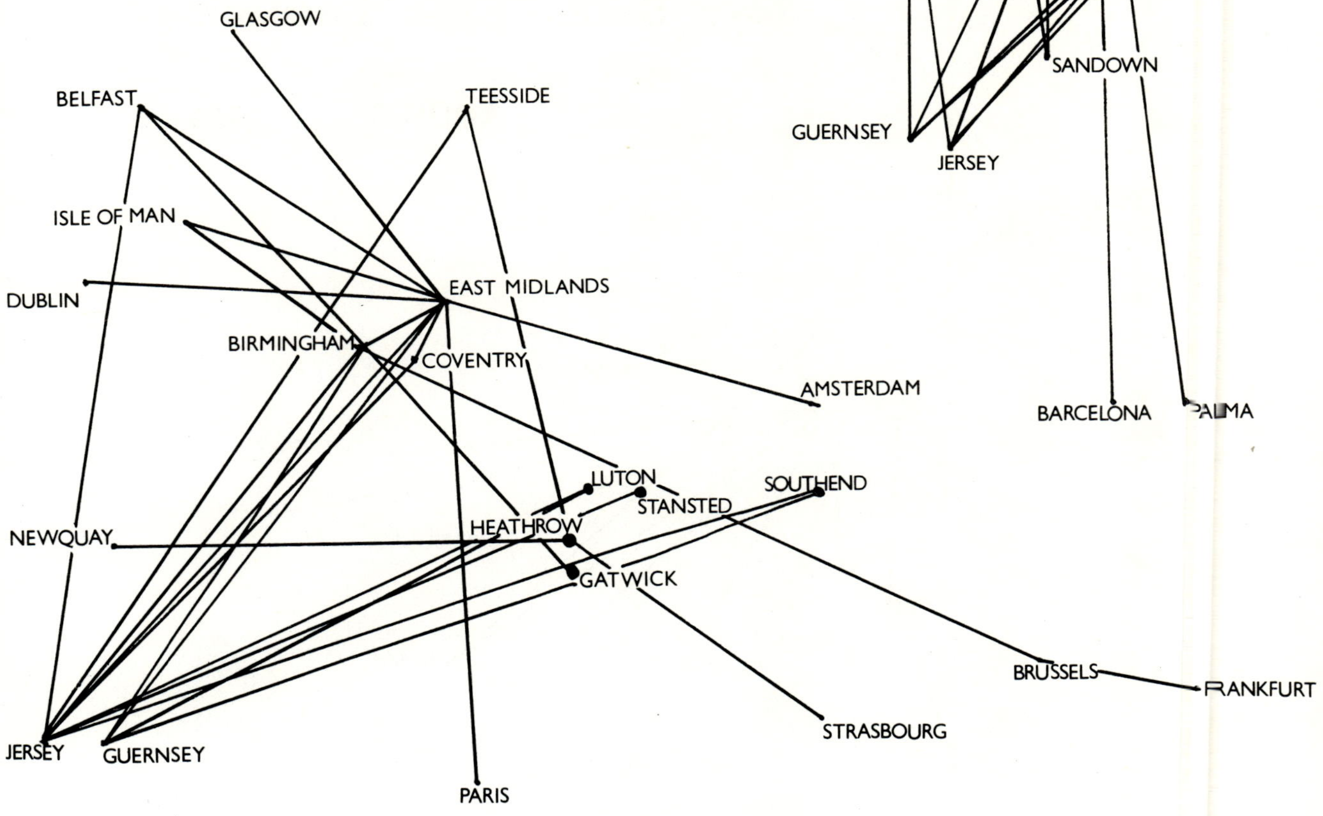

British Midland Airways 1975